PIC32 Microcontrollers and the Digilent chipKIT

Introductory to Advanced Projects

Dogan Ibrahim

ELSEVIER

AMSTERDAM • BOSTON • HEIDELBERG • LONDON
NEW YORK • OXFORD • PARIS • SAN DIEGO
SAN FRANCISCO • SINGAPORE • SYDNEY • TOKYO
Newnes is an Imprint of Elsevier

Newnes

Newnes is an imprint of Elsevier
The Boulevard, Langford Lane, Kidlington, Oxford OX5 1GB, UK
225 Wyman Street, Waltham, MA 02451, USA

First edition 2015

British Library Cataloguing in Publication Data
A catalogue record for this book is available from the British Library

Library of Congress Cataloging-in-Publication Data
A catalog record for this book is available from the Library of Congress

ISBN: 978-0-08-099934-0

For information on all Newnes publications
visit our web site at http://store.elsevier.com/

Printed and bound in the United States of America
Typeset by Thomson Digital

14 15 16 17 18 10 9 8 7 6 5 4 3 2 1

Contents

Preface

A microcontroller is a single-chip microprocessor system that contains data and program memory, serial and parallel I/O, timers, and external and internal interrupts, all integrated into a single chip that can be purchased for as little as $2.00. About 40% of microcontroller applications are in office automation, such as PCs, laser printers, fax machines, intelligent telephones, and so forth. About one-third of microcontrollers are found in consumer electronic goods. Products such as CD players, hi-fi equipment, video games, washing machines, cookers, and so on fall into this category. The communications market, automotive market, and the military share the rest of the application areas.

This book is written for students, for practising engineers, and for hobbyists who want to learn more about the programming and applications of PIC 32-bit series of microcontrollers. It has been written with the assumption that the reader has taken a course on digital logic design, and has been exposed to writing programs using at least one high-level programming language. Knowledge of the C programming language will be useful. Also, familiarity with at least one member of the PIC series of microcontrollers (e.g., PIC16 or PIC18) will be an advantage. The knowledge of assembly language programming is not required because all the projects in the book are based on using the C (and C++) language.

chipKIT is a series of 32-bit PIC microcontroller-based development boards designed and manufactured by Digilent Inc. (www.digilentinc.com). There are many such development boards in the series, starting from the basic chipKIT Uno32 and chipKIT MX3 to more advanced chipKIT Pro MX4, chip KIT Pro MX7, and so on. These boards are complete 32-bit microcontroller development systems compatible with the MPIDE and the MPLAB IDE integrated development environments. The boards are supported by a large number of plug-in peripheral interface modules called Pmods. Some examples of Pmod modules are: LED arrays, seven-segment display, push-button switches, keypad, LCD, OLED, temperature sensor, GPS, stepper motor controller, and many more. The chipKIT PIC32-based systems are compatible with many existing Arduino® code examples, reference material, and other resources, thus making the programming easy. In this book, many tested and working projects are given based on the chipKIT MX3 development board and the MPIDE integrated development environment.

Chapter 1 presents the basic features of microcontrollers and gives example of a simple microcontroller-based fluid-level control system.

Chapter 2 provides a review of the PIC32 series of 32-bit microcontrollers. Various features of these microcontrollers are described in detail. The highly popular PIC32MX360F512L microcontroller has been chosen as an example 32-bit microcontroller.

Chapter 3 is about the 32-bit microcontroller development tools. The basic features of various development boards available in the market are given in this chapter.

Chapter 4 discusses the features of the popular chipKIT MX3 (formerly known as the Cerebot MX3cK) 32-bit development board in detail. This development board is used in the projects in this book.

Chapter 5 is about the MPIDE integrated development environment. The chapter discusses various features of this IDE and gives programming examples.

Chapter 6 describes the commonly used microcontroller program development tools. Flow charts and the program description language (PDL) are explained in this chapter with examples.

Chapter 7 provides projects using the chipKIT MX3 development board with the MPIDE integrated development environment. Many tested and working projects are given in this chapter. The following are given for each project:

- Project title
- Project description
- Project hardware
- Project PDL
- Complete project program listing
- Full description of the program
- Comments for future development (where necessary)

Finally, the Appendix describes how the MPLAB IDE can be used in developing applications with the chipKIT boards. chipKIT Pro MX7 (formerly known as the Cerebot MX7cK) development board is taken as an example, and a simple project is developed to illustrate the basic steps.

Dogan Ibrahim
London

Acknowledgments

The following material is reproduced in this book with the kind permission of the respective copyright holders and may not be reprinted, or reproduced in any way, without their prior consent.

Figures 2.1–2.5, 2.8, and 2.15–2.35 are taken from Microchip Technology Inc. Data Sheet PIC32MX3XX/4XX (DS61143E). Figures 3.1–3.5 are taken from the web site of Microchip Technology Inc.

Figures 3.6–3.11 and Pmod peripheral interface module pictures are taken from the web site of Digilent Inc.

Figures 3.12–3.16, 3.19, and 3.20 are taken from the web site of mikroElektronika.

Figures 3.17 and 3.18 are taken from the web site of Olimex.

I would like to thank my son Ahmet Ibrahim who has helped in the construction and testing of the projects in this book.

PIC®, PICSTART®, and MPLAB® are all trademarks of Microchip Technology Inc.

CHAPTER 1

Microcomputer Systems

Chapter Outline

1.1 Introduction

The term microcomputer is used to describe a system that includes a minimum of a microprocessor, program memory, data memory, and input–output (I/O). Some microcomputer systems include additional components such as timers, counters, analogue-to-digital converters (ADCs), and so on. Thus, a microcomputer system can be anything from a large computer having hard disks, floppy disks, and printers to a single-chip embedded controller.

In this book, we are going to consider only the type of microcomputers that consists of a single silicon chip. Such microcomputer systems are also called microcontrollers, and they are used in many household goods such as microwave ovens, TV remote control units, cookers, hi-fi equipment, CD players, personal computers, fridges, etc. There are a large number of microcontrollers available in the market. In this book, we shall be looking at the programming and system design using the 32-bit *programmable interface controller* (PIC) series of microcontrollers manufactured by Microchip Technology Inc.

1.2 Microcontroller Systems

A microcontroller is a single-chip computer. *Micro* suggests that the device is small, and *controller* suggests that the device can be used in control applications. Another term used for microcontrollers is *embedded controller*, since most of the microcontrollers are built into (or embedded in) the devices they control.

A microprocessor differs from a microcontroller in many ways. The main difference is that a microprocessor requires several other components for its operation, such as program memory and data memory, I/O devices, and external clock circuit. A microcontroller, on the other hand, has all the support chips incorporated inside the same chip. All microcontrollers operate on a set of instructions (or the user program) stored in their memory. A microcontroller fetches the instructions from its program memory one by one, decodes these instructions, and then carries out the required operations.

Microcontrollers have traditionally been programmed using the assembly language of the target device. Although the assembly language is fast, it has several disadvantages. An assembly program consists of mnemonics, and it is difficult to learn and maintain a program written using the assembly language. Also, microcontrollers manufactured by different firms have different assembly languages and the user is required to learn a new language every time a new microcontroller is to be used. Microcontrollers can also be programmed using a high-level language, such as BASIC, PASCAL, and C. High-level languages have the advantage that it is much easier to learn a high-level language than an assembler. Also, very large and complex programs can easily be developed using a high-level language.

In general, a single chip is all that is required to have a running microcontroller system. In practical applications, additional components may be required to allow a microcomputer to interface to its environment. With the advent of the PIC family of microcontrollers, the development time of an electronic project has reduced to several hours.

Basically, a microcomputer (or microcontroller) executes a user program that is loaded in its program memory. Under the control of this program, data is received from external devices (inputs), manipulated, and then sent to external devices (outputs). For example, in a microcontroller-based fluid level control system, the fluid level is read by the microcomputer via a level sensor device and the microcontroller attempts to control the fluid level at the required value. If the fluid level is low, the microcomputer operates a pump to draw more fluid from the reservoir in order to keep the fluid at the required level. Figure 1.1 shows the block diagram of our simple fluid level control system.

The system shown in Figure 1.1 is a very simplified fluid level control system. In a more sophisticated system, we may have a keypad to set the required fluid level and an LCD

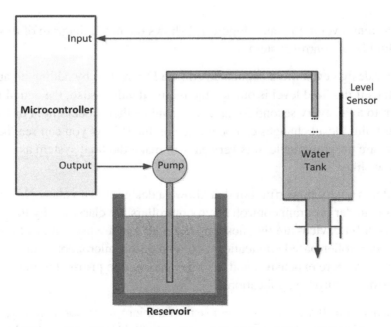

Figure 1.1: Microcontroller-Based Fluid Level Control System

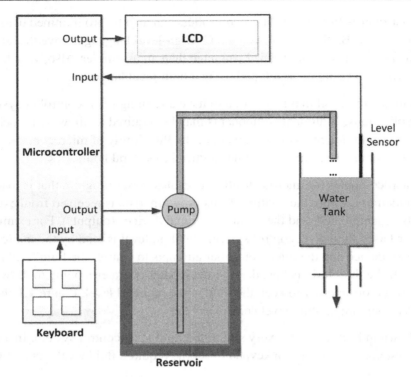

Figure 1.2: Fluid Level Control System With a Keypad and LCD

to display the current level in the tank. Figure 1.2 shows the block diagram of this more sophisticated fluid level control system.

We can make our design even more sophisticated (see Figure 1.3) by adding an audible alarm to inform us if the fluid level is outside the required value. Also, the actual level at any time can be sent to a PC every second for archiving and further processing. For example, a graph of the daily fluid level changes can be plotted on the PC. As you can see, because the microcontrollers are programmable, it is very easy to make the final system as simple or as complicated as we like.

A microcontroller is a very powerful tool that allows a designer to create sophisticated I/O data manipulation under program control. Microcontrollers are classified by the number of bits they process. 8-bit devices are the most popular ones and are used in most low-cost, low-speed microcontroller-based applications. 16- and 32-bit microcontrollers are much more powerful, but usually more expensive, and their use may not be justified in many small- to medium-sized general-purpose applications.

The simplest microcontroller architecture consists of a microprocessor, memory, and I/O. The microprocessor consists of a central processing unit (CPU) and the control unit (CU).

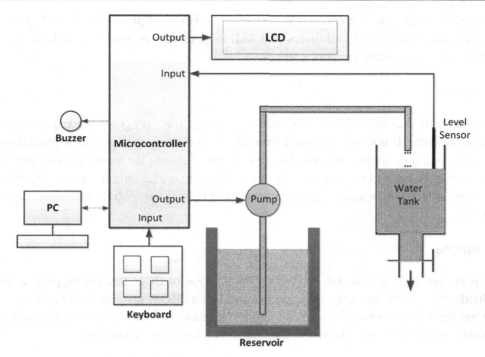

Figure 1.3: More Sophisticated Fluid Level Controller

The CPU is the brain of the microcontroller, and this is where all of the arithmetic and logic operations are performed. The CU controls the internal operations of the microprocessor and sends out control signals to other parts of the microcontroller to carry out the required instructions.

Memory is an important part of a microcontroller system. Depending on the type used, we can classify memories into two groups: program memory and data memory. Program memory stores the programs written by the programmer, and this memory is usually nonvolatile, that is, data is not lost after the removal of power. Data memory is where the temporary data used in a program is stored, and this memory is usually volatile, that is, data is lost after the removal of power.

There are basically six types of memories, summarised as follows.

1.2.1 RAM

RAM means random access memory. It is a general-purpose memory that usually stores the user data in a program. RAM memory is volatile in the sense that it cannot retain data in the absence of power, that is, data is lost after the removal of power. Most microcontrollers have some amount of internal RAM. Several kilobytes is a common amount, although some microcontrollers have much more, and some have less. For example, the PIC32MX460F512L 32-bit microcontroller has 512 kilobytes of RAM. In general, it is possible to extend the memory by adding external memory chips.

1.2.2 ROM

ROM is read-only memory. This type of memory usually holds program or fixed user data. ROM is nonvolatile. If power is removed from ROM and then reapplied, the original data will still be there. ROM memories are programmed at factory during the manufacturing process, and their contents cannot be changed by the user. They are useful only if we have developed a program and wish to order several thousand copies of it, or if we wish to store some configuration data.

1.2.3 PROM

PROM is programmable read-only memory. This is a type of ROM that can be programmed in the field, often by the end user, using a device called a PROM programmer. Once a PROM has been programmed, its contents cannot be changed. PROMs are usually used in low-production applications where only several such memories are required.

1.2.4 EPROM

EPROM is erasable programmable read-only memory. This is similar to ROM, but the EPROM can be programmed using a suitable programming device. EPROM memories have a small clear glass window on top of the chip where the data can be erased under strong ultraviolet light. Once the memory is programmed, the window can be covered with dark tape to prevent accidental erasure of the data. An EPROM memory must be erased before it can be reprogrammed. Many development versions of microcontrollers are manufactured with EPROM memories where the user program can be stored. These memories are erased and reprogrammed until the user is satisfied with the program. Some versions of EPROMs, known as one-time programmable (OTP), can be programmed using a suitable programmer device, but these memories cannot be erased. OTP memories cost much less than the EPROMs. OTP is useful after a project has been developed completely and it is required to make many copies of the program memory.

1.2.5 EEPROM

EEPROM is electrically erasable programmable read-only memory, which is a nonvolatile memory. These memories can be erased and also be reprogrammed using suitable programming devices. EEPROMs are used to save configuration information, maximum and minimum values, identification data, etc.

1.2.6 Flash EEPROM

This is another version of EEPROM-type memory. This memory has become popular in microcontroller applications and is generally used to store the user program. Flash EEPROM is nonvolatile and is usually very fast. The data can be erased and then reprogrammed using a suitable programming device. These memories can also be programmed without removing them from their circuits. Some microcontrollers have only 1k flash EEPROM, while some others have 32k or more.

1.3 Microcontroller Features

Microcontrollers from different manufacturers have different architectures and different capabilities. Some may suit a particular application, while others may be totally unsuitable for the same application. The hardware features of microcontrollers in general are described in this section.

1.3.1 Supply Voltage

Most microcontrollers operate with the standard logic voltage of +5 V. Some microcontrollers can operate at as low as +2.7 V, and some will tolerate +6 V without any problems. You should check the manufacturers' data sheets about the allowed limits of the power supply voltage. For example, PIC32MX460F512L 32-bit microcontrollers can operate with a power supply of +2.3 to +3.6 V.

A voltage regulator circuit is usually used to obtain the required power supply voltage when the device is to be operated from a mains adaptor or batteries. For example, a 5 V regulator may be required if the microcontroller and peripheral devices operate from a +5 or +3.3 V supply and a 9 V battery is to be used as the power supply.

1.3.2 The Clock

All microcontrollers require a clock (or an oscillator) to operate. The clock is usually provided by connecting external timing devices to the microcontroller. Most microcontrollers will generate clock signals when a crystal and two small capacitors are connected. Some will operate with resonators or external resistor–capacitor pairs. Some microcontrollers have built-in timing circuits, and they do not require any external timing components. If your application is not time-sensitive, you should use external or internal (if available) resistor–capacitor timing components for simplicity and low cost.

An instruction is executed by fetching it from the memory and then decoding it. This usually takes several clock cycles and is known as the *instruction cycle*. PIC32 series of microcontrollers can operate with clock frequencies up to 80 MHz.

1.3.3 Timers

Timers are important parts of any microcontroller. A timer is basically a counter that is driven either from an external clock pulse or from the internal oscillator of the microcontroller. A PIC32 microcontroller can have 16- or 32-bit wide timers (two 16-bit timers are combined to create a 32-bit timer). Data can be loaded into a timer under program control, and the timer can be stopped or started by program control. Most timers can be configured to generate an interrupt when they reach a certain count (usually when they overflow). The interrupt can be used by the user program to carry out accurate timing-related operations inside the microcontroller.

Some microcontrollers offer capture and compare facilities where a timer value can be read when an external event occurs, or the timer value can be compared with a preset value and an interrupt can be generated when this value is reached. PIC32 microcontrollers can have up to five capture inputs.

1.3.4 Watchdog

Most microcontrollers have at least one watchdog facility. The watchdog is basically a timer that is refreshed by the user program and a reset occurs if the program fails to refresh the watchdog. The watchdog timer is used to detect a system problem, such as the program being in an endless loop. A watchdog is a safety feature that prevents runaway software and stops the microcontroller from executing meaningless and unwanted code. Watchdog facilities are commonly used in real-time systems for safety where it may be required to regularly check the successful termination of one or more activities.

1.3.5 Reset Input

A reset input is used to reset a microcontroller externally. Resetting puts the microcontroller into a known state such that the program execution starts from a known address. An external reset action is usually achieved by connecting a push-button switch to the reset input such that the microcontroller can be reset when the switch is pressed.

1.3.6 Interrupts

Interrupts are very important concepts in microcontrollers. An interrupt causes the microcontroller to respond to external and internal (e.g., a timer) events very quickly. When an interrupt occurs, the microcontroller leaves its normal flow of program execution and jumps to a special part of the program, known as the *interrupt service routine* (ISR). The program code inside the ISR is executed, and on return from the ISR the program resumes its normal flow of execution.

The ISR starts from a fixed address of the program memory. This address is also known as the *interrupt vector address*. Some microcontrollers with multi-interrupt features have just one interrupt vector address, while some others have unique interrupt vector addresses, one for each interrupt source. Interrupts can be nested such that a new interrupt can suspend the execution of another interrupt. Another important feature of a microcontroller with multi-interrupt capability is that different interrupt sources can be given different levels of priority.

1.3.7 Brown-Out Detector

Brown-out detectors are also common in many microcontrollers, and they reset a microcontroller if the supply voltage falls below a nominal value. They are safety features, and they can be employed to prevent unpredictable operation at low voltages, especially to protect the contents of EEPROM-type memories.

1.3.8 Analogue-to-Digital Converter

An ADC is used to convert an analogue signal such as voltage to a digital form so that it can be read and processed by a microcontroller. Some microcontrollers have built-in ADC converters. It is also possible to connect an external ADC converter to any type of microcontroller. PIC32 microcontroller ADC converters are usually 10-bit wide, having 1024 quantisation levels. Most PIC microcontrollers with ADC features have multiplexed ADC converters where more than one analogue input channel is provided. For example, PIC32MX460F512L microcontroller has 16 channels of ADC converters, each 10-bit wide.

The ADC conversion process must be started by the user program, and it may take several tens of microseconds for a conversion to complete. ADC converters usually generate interrupts when a conversion is complete so that the user program can read the converted data quickly and efficiently.

ADC converters are very useful in control and monitoring applications since most sensors (e.g., temperature sensor, pressure sensor, force sensor) produce analogue output voltages.

1.3.9 Serial Input–Output

Serial communication (also called RS232 communication) enables a microcontroller to be connected to another microcontroller or to a PC using a serial cable. Some microcontrollers have built-in hardware called universal synchronous–asynchronous receiver–Transmitter (UART) to implement a serial communication interface. The baud rate (bits per second) and the data format can usually be selected by the user program. If any serial I/O hardware is not provided, it is easy to develop software to implement serial data communication using any I/O pin of a microcontroller. PIC32MX460F512L microcontroller provides two UART modules, each with RS232-, RS485-, LIN-, or IrDA-compatible interfaces.

In addition, two Serial Peripheral Interface (SPI) and two Integrated InterConnect (I²C) hardware bus interfaces are available on the PIC32MX460F512L microcontroller.

1.3.10 EEPROM Data Memory

EEPROM-type data memory is also very common in many microcontrollers. The advantage of an EEPROM memory is that the programmer can store nonvolatile data in such a memory, and can also change this data whenever required. For example, in a temperature monitoring application, the maximum and the minimum temperature readings can be stored in an EEPROM memory. Then, if the power supply is removed for whatever reason, the values of the latest readings will still be available in the EEPROM memory.

1.3.11 LCD Drivers

LCD drivers enable a microcontroller to be connected to an external LCD display directly. These drivers are not common since most of the functions provided by them can easily be implemented in software.

1.3.12 Analogue Comparator

Analogue comparators are used where it is required to compare two analogue voltages. These modules are not available in low-end PIC microcontrollers. PIC32MX460F512L microcontrollers have two built-in analogue comparators.

1.3.13 Real-Time Clock

Real-time clock enables a microcontroller to have absolute date and time information continuously. Built-in real-time clocks are not common in most microcontrollers since they can easily be implemented by either using an external dedicated real-time clock chip or writing a program. For example, PIC32MX460F512L microcontroller has built-in real-time clock and calendar module.

1.3.14 Sleep Mode

Some microcontrollers offer built-in sleep modes where executing this instruction puts the microcontroller into a mode where the internal oscillator is stopped and the power consumption is reduced to an extremely low level. The main reason of using the sleep mode is to conserve the battery power when the microcontroller is not doing anything useful. The microcontroller usually wakes up from the sleep mode by external reset or by a watchdog time-out.

1.3.15 Power-On Reset

Some microcontrollers have built-in power-on reset circuits that keep the microcontroller in reset state until all the internal circuitry has been initialised. This feature is very useful as it starts the microcontroller from a known state on power-up. An external reset can also be provided where the microcontroller can be reset when an external button is pressed.

1.3.16 Low-Power Operation

Low-power operation is especially important in portable applications where the microcontroller-based equipment is operated from batteries, and very long battery life is a main requirement. Some microcontrollers can operate with less than 2 mA with 5 V supply, and around 15 μA at 3 V supply.

1.3.17 Current Sink/Source Capability

This is important if the microcontroller is to be connected to an external device that may draw large current for its operation. PIC32 microcontrollers can source and sink 18 mA of current from each output port pin. This current is usually sufficient to drive LEDs, small lamps, buzzers, small relays, etc. The current capability can be increased by connecting external transistor switching circuits or relays to the output port pins.

1.3.18 USB Interface

USB is currently a very popular computer interface specification used to connect various peripheral devices to computers and microcontrollers. PIC32 microcontrollers normally provide built-in USB modules.

1.3.19 Motor Control Interface

Some PIC microcontrollers, for example, PIC18F2x31, provide motor control interface.

1.3.20 CAN Interface

CAN bus is a very popular bus system used mainly in automation applications. Some PIC18F series of microcontrollers (e.g., PIC18F4680) provide CAN interface capabilities.

1.3.21 Ethernet Interface

Some PIC microcontrollers (e.g., PIC18F97J60) provide Ethernet interface capabilities. Such microcontrollers can easily be used in network-based applications.

1.3.22 ZigBee Interface

ZigBee is an interface similar to Bluetooth and is used in low-cost wireless home automation applications. Some PIC series of microcontrollers provide ZigBee interface capabilities, making the design of such wireless systems very easy.

1.3.23 Multiply and Divide Hardware

PIC32 microcontrollers have built-in hardware for fast multiplication and division operations.

1.3.24 Operating Temperature

It is important to know the operating temperature range of a microcontroller chip before a project is developed. PIC32 microcontrollers can operate in the temperature range −40 to +105°C.

1.3.25 Pulse Width Modulated (PWM) Outputs

Most microcontrollers provide PWM outputs for driving analogue devices such as motors, lamps, etc. The PWM is usually a separate module and runs in hardware, independent of the CPU.

1.3.26 JTAG Interface

Some high-end microcontrollers (e.g., PIC32MX460F512L) have built-in JTAG interfaces for enhanced debugging interface.

1.3.27 Package Size

It is sometimes important to know the package size of a microcontroller chip before a microcontroller is chosen for a particular project. Low-end microcontrollers are usually packaged in 18-, 28-, or 40-pin DIL packages. High-end microcontrollers, for example, PIC32MX-460F512L, are housed in a 100-pin TQFP-type package.

1.3.28 DMA

Some high-end microcontrollers have built-in direct memory access (DMA) channels that can be used to transfer large amounts of data between different devices without the intervention of the CPU. For example, PIC32MX460F512L microcontroller has four DMA channels.

1.4 Microcontroller Architectures

Usually two types of architectures are used in microcontrollers (see Figure 1.4): *Von Neumann* architecture and *Harvard* architecture. Von Neumann architecture is used by a large percentage of microcontrollers, and here all memory space is on the same bus and instruction

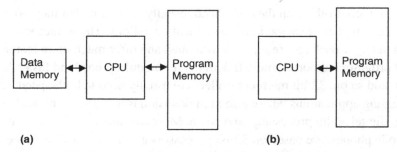

Figure 1.4: (a) Harvard and (b) Von Neumann Architectures

and data use the same bus. In the Harvard architecture (used by most PIC microcontrollers), code and data are on separate buses, and this allows the code and data to be fetched simultaneously, resulting in an improved performance.

1.4.1 RISC and CISC

Reduced instruction set computer (RISC) and complex instruction set computer (CISC) refer to the instruction set of a microcontroller. In an 8-bit RISC microcontroller, data is 8-bit wide but the instruction words are more than 8-bit wide (usually 12-, 14-, or 16-bit) and the instructions occupy one word in the program memory. Thus, the instructions are fetched and executed in one cycle, resulting in an improved performance.

In a CISC microcontroller, both data and instructions are 8-bit wide. CISC microcontrollers usually have over 200 instructions. Data and code are on the same bus and cannot be fetched simultaneously.

1.5 8, 16, or 32 Bits?

People are usually confused for making a decision between 8, 16, or 32 bits of microcontrollers. It is important to realise that the number of bits just refers to the width of the data handled by the processor. This number actually limits the precision of mathematical operations carried out by the CPU (although it is possible to emulate higher-order mathematics in software or by using special hardware).

In general, 8-bit microcontrollers have been around since the first days of the microcontroller development. They are cheap, easy to use (only small package size), low speed, and can be used in most general-purpose control and data manipulation operations. For example, it is still very efficient to design low- to medium-speed control systems (e.g., temperature control, fluid level control, or robotics applications) using 8-bit microcontrollers. In such applications, low cost is more important than high speed. Many commercial and industrial applications fall into this category and can easily be designed using standard 8-bit microcontrollers.

16- and 32-bit microcontrollers, on the other hand, usually cost more, but they offer much higher speeds, and much higher precision in mathematical operations. These microcontrollers are usually housed in larger packages (e.g., 64 or 100 pins) and offer much more features, such as larger data and program memories, more timer/counter modules, more and faster A/D channels, more I/O ports, and so on. 32-bit microcontrollers are usually used in high-speed, real-time digital signal processing applications where also high precision is a requirement, such as digital image processing, digital audio processing, and so on. Most consumer products, such as electronic games and mobile phones, are based on 32-bit processors as they demand high-speed real-time operation with colour graphical displays and with touchscreen panels. Other high-speed applications such as video capturing, image filtering, video editing, video streaming, speech recognition, and speech processing all require very fast 32-bit processors with lots of data and program memories, and very high precision while implementing the digital signal processing algorithms.

This book is about 32-bit PIC microcontrollers. We shall be seeing the basic architectures and features of these microcontrollers. In addition, many working projects will be given in the book to illustrate how these microcontrollers can be programmed and used in real applications. The Cerebot MX3cK Development Board (has been renamed to chipKIT MX3) will be used in the projects.

1.6 Summary

This chapter has given an introduction to the microprocessor and microcontroller systems. The basic building blocks of microcontrollers have been described briefly with reference to various types of memories used in microcontroller systems.

1.7 Exercises

1. What is a microcontroller? What is a microprocessor? Explain the main differences between a microprocessor and a microcontroller.
2. Give some examples of applications of microcontrollers around you.
3. Where would you use an EPROM memory?
4. Where would you use a RAM memory?
5. Explain what types of memories are usually used in microcontrollers.
6. What is an input–output port?
7. What is an analogue-to-digital converter? Give an example use for this converter.
8. Explain why a watchdog timer could be useful in a real-time system.
9. What is serial input–output? Where would you use serial communication?
10. Why is the current sinking/sourcing important in the specification of an output port pin?
11. What is an interrupt? Explain what happens when an interrupt is recognised by a microcontroller.
12. Why is brown-out detection important in real-time systems?
13. Explain the differences between a RISC-based microcontroller and a CISC-based microcontroller. What type of microcontroller is PIC?

CHAPTER 2

PIC32 Microcontroller Series

Chapter Outline

PIC32 is a 32-bit family of general-purpose, high-performance microcontrollers manufactured by Microchip Technology Inc. Looking at the PIC microcontroller development history, based on their performance, we can divide the PIC microcontroller families into low-performance, low-to-medium–performance, medium-performance, and high-performance devices.

Low-performance PIC microcontrollers consist of the basic 8-bit PIC10 and PIC16 series of devices that have been around for over a decade. These devices are excellent general-purpose low-speed microcontrollers that have been used successfully in thousands of applications worldwide.

The PIC18 series of microcontrollers were then introduced by Microchip Technology Inc. as low-to-medium–performance devices for use in high–pin count, high-density, complex applications requiring large number of I/O ports, large program and data memories, and supporting complex communication protocols such as CAN, USB, TCP/IP, or ZigBee. Although these devices are also based on 8-bit architecture, they offer higher speeds, from DC to 40 MHz, with a performance rating of up to 10 MIPS.

The PIC24 series of microcontrollers are based on 16-bit architecture and have been intro-duced as medium-performance devices to be used in applications requiring high compatibility with lower-performance PIC microcontroller families, and at the same time offering higher throughput and complex instruction sets. These microcontrollers have been used in many real-time applications such as digital signal processing, automatic control, speech and image processing, and so on, where higher accuracy than 8 bits is required, and at the same time higher speed is the main requirement.

PIC32 microcontroller family has been developed for high-performance, general-purpose microcontroller applications. The family offers 80 MIPS performance with a wide range of on-chip peripherals, large data and program memories, large number of I/O ports, and an architecture designed for high-speed real-time applications. PIC32 microcontrollers can be used in real-time applications requiring high throughput. Some of the application areas are digital signal processing, digital automatic control, real-time games, and fast communication. The chip employs the industry standard M4K MIPS32 core from MIPS Technologies Inc. PIC32 family offers programming interface similar to other PIC microcontroller families, thus making the programming an easy task if the programmer is already familiar with the basic PIC microcontroller architecture. PIC32 microcontrollers are pin-to-pin compatible with most members of the PIC24 family of 16-bit microcontrollers; thus, the migration from 16- to 32-bit operation should be relatively easy.

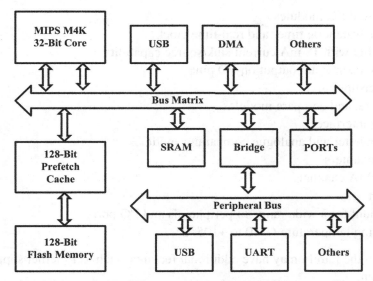

Figure 2.1: Simplified Architecture of the PIC32 Microcontroller Family

Figure 2.1 shows a simplified architectural overview of the PIC32 microcontrollers. At the heart of the microcontroller is a 32-bit M4K MIPS32 core processor that connects to the rest of the chip via a *bus matrix* and a *peripheral bus*. The bus matrix runs at the same speed as the core processor and connects various high-speed modules such as the USB, DMA, memory, ports, and so on. The peripheral bus can be programmed to run at slower speeds, and it connects to slower modules such as A/D converter, UART, SPI, Real-Time Clock and Calendar (RTCC), and so on.

The core processor has the following features:

- 80 MHz clock speed
- 32-Bit address bus and 32-bit data bus
- Five-stage pipelining
- Single-cycle ALU
- Single-cycle multiply and high-speed divide module
- 2 × 32 register files

Other important features of the chip are the following:

- 2.3–3.6 V operation
- Up to 512k flash program memory
- Up to 32k SRAM data memory
- Internal oscillators
- Multiple interrupt vectors

- UART, SPI, and I²C modules
- Configurable watchdog timer and real-time clock
- High-speed I/O with 18 mA current sink/source capability
- Configurable open-drain output on I/O pins
- External interrupt pins
- PWM, capture, and compare modules
- JTAG debug interface
- Fast A/D converter and analogue comparator modules
- Timers and counters
- Hardware DMA channels
- USB support
- Large pin count for a wide range of peripherals and I/O ports
- Wide operating temperature (-40 to $+105°C$)

Different chips in the family may have additional features, such as Ethernet support, CAN bus support, and so on.

Perhaps the best way of learning the PIC32 microcontroller family architecture is to look at a typical processor in the family in greater detail. In this book, the Digilent Cerebot MX-3cK Development Board (has been renamed to chipKIT MX3) will be used. This board is equipped with the PIC32MX320F128H microcontroller (from the PIC32MX3XX/4XX family). In this chapter, we shall be looking at the architecture of the PIC32MX360F512L microcontroller from the same family. The architectures of other members of the PIC32 microcontroller family are very similar to the chosen one and should not be too difficult to learn them.

2.1 The PIC32MX360F512L Architecture

The PIC32MX360F512L is a typical PIC32 microcontroller, belonging to the family PIC32MX3XX/4XX. This microcontroller has the following features:

- A 100-pin (TQFP) package
- Up to 80 MHz clock speed
- 512k flash program memory (+12k boot flash memory)
- 32k data memory
- $5 \times$ 16-bit timers/counters
- $5 \times$ capture inputs
- $5 \times$ compare/PWM outputs
- Four programmable DMA channels
- $2 \times$ UARTs (supporting RS232, RS485, LIN bus, and iRDA)
- $2 \times$ SPI bus modules
- $2 \times$ I²C bus modules
- RTCC module
- 8 MHz and 32 kHz internal clocks

- 16 × A/D channels (10 bits)
- 2 × comparator modules
- Four-wire JTAG interface

PIC32MX360F512L is available in a 100-pin TQFP package as shown in Figure 2.2. The pins that are +5 V–tolerant are shown in bold.

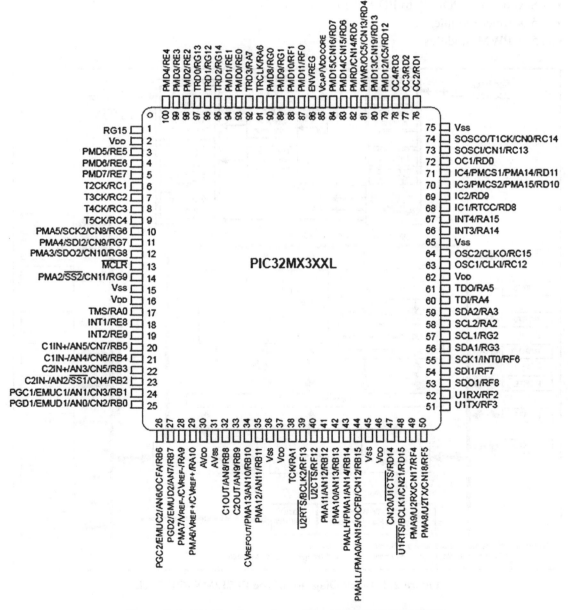

Figure 2.2: Pin Configuration of the PIC32MX360F512L

Figure 2.3 shows the internal block diagram of the PIC32MX360F512L microcontroller. In the middle of the diagram, we can see the MIP32 M4K processor. The 32-bit data memory is directly connected to the processor via the bus matrix, offering up to 4 GB addressing space. The 1298-bit flash program memory is also connected to the bus matrix via a 32-bit prefetch module. The 32-bit peripheral bridge connects the bus matrix and the processor to the peripheral modules. The peripheral modules consist of the following:

- Seven ports (PORTA to PORTG)
- 5 × timer modules
- 5 × PWM modules

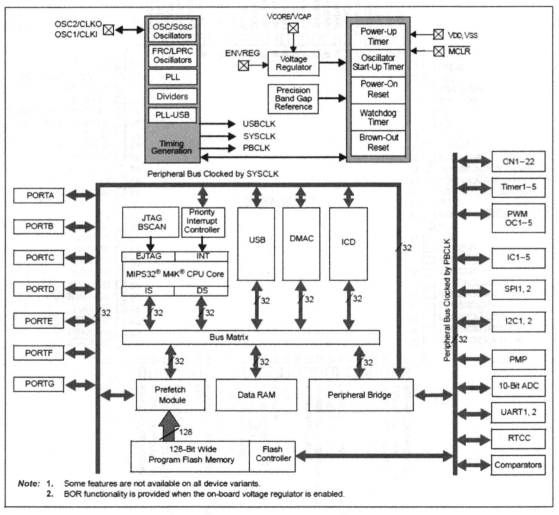

Figure 2.3: Block Diagram of the PIC32MX360F512L

- 2 × SPI bus modules
- 2 × I²C modules
- 10-Bit ADC module
- 2 × UART modules
- RTCC module
- Comparators
- Change notification inputs

The system clock and peripheral bus clock are provided by the Timing Generation module, which consists of the following:

- Oscillators
- PLL module
- Clock dividers

The Timing Generation module additionally provides clock to the following modules:

- Power-up timer
- Oscillator start-up timer
- Power-on reset (POR)
- Watchdog timer
- Brown-out reset (BOR)

The functions of the pins are summarised in Table 2.1.

2.1.1 The Memory

Figure 2.4 shows the memory structure of the PIC32MX3XX/4XX microcontrollers. The memory structure may look complicated initially, but the explanation given in this section should make it simple and easy to understand.

As can be seen from the figure, two address spaces are implemented: *virtual* and *physical*. All hardware resources, such as data memory, program memory, and DMA transfers, are handled by the physical addresses. If we wish to access the memory independent of the CPU (such as the case in DMA), then we must use the physical addresses.

Virtual addresses are important as they are used exclusively by the CPU to fetch and execute instructions. In normal programming, we are only interested with the virtual memory addresses. These addresses are translated into physical addresses by a Fixed Mapping Translation (FMT) unit inside the processor. The translation process is simply a bitwise AND of the virtual address spaces with fixed number 0x1FFFFFFF. Details of the physical address space are available in the *PIC32 Family Reference Manual*, or in the individual microcontroller data sheets.

Table 2.1: Pin descriptions.

Pin Name	Description
AN0–AN15	Analogue inputs
CN0–CN21	Change notification inputs
IC1–IC5	Capture inputs
OC1–OC5	Output compare outputs
INT0–INT4	External interrupt inputs
RA0–RA15	PORTA pins
RB0–RB15	PORTB pins
RC1–RC15	PORTC pins
RD0–RD15	PORTD pins
RE0–RE9	PORTE pins
RF0–RF13	PORTF pins
RG0–RG15	PORTG pins
T1CK–T5CK	Timer external clock inputs
U1CTS/U1RTS/U1RX/U1TX	UART1 pins
U2CTS/U2RTS/U2RX/U2TX	UART2 pins
SDI1/SDO1/SS1/SCK1	SPI1 pins
SDI2/SDO2/SS2/SCK2	SPI2 pins
SCL1/SDA1	I²C1 pins
SCL2/SDA2	I²C2 pins
TMS/TCK/TDI/TDO	JTAG pins
RTCC	Real time clock alarm output
CVREF-/CVREF+/CVREFOUT	Comparator voltage reference pins
C1IN-/C1IN+/C1OUT	Comparator 1 input-outputs
C2IN-/C2IN+/C2OUT	Comparator 2 input-outputs
PMA0–PMA15	Parallel Master Port address bits
PMCS1/PMCS2	Parallel Master Port chip select pins
PMD0–PMD15	Parallel Master Port data pins
PMRD/PMWR/PMALL/PMALH	Parallel Master Port control pins
OCFA/OCFB	Output compare fault inputs
TRCLK	Trace clock
TRD0–TRD3	Trace data pins
PGED1/PGEC1/PGED2/PGEC2	Programming/debugging pins
MCLR	Master Clear pin
AVDD/AVSS/VREF+/VREF-	ADC pins
VDD/VSS	Microcontroller supply pins
OSC1/OSC2/CLK0/CLK1/SOSC1/SOSCO	Oscillator input-outputs

The entire virtual memory address space is 4 GB and is divided into two primary regions: user mode segment and kernel mode segment. The lower 2 GB of address space forms the user mode segment (called Useg/Kuseg). The upper 2 GB of virtual address space forms the kernel mode segment. The kernel address space has been designed to be used by the operating system, while the user address space holds a user program that runs under the operating system. This is for safety, as a program in the user address space cannot access the memory space of the kernel address space. But programs in the kernel can access the user address

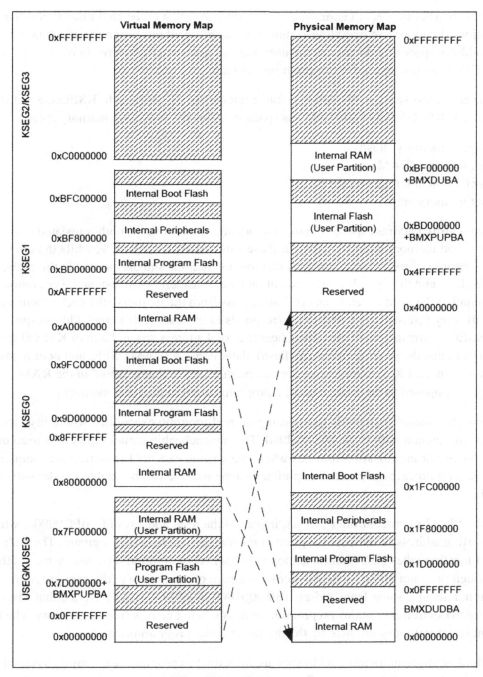

Figure 2.4: The Memory Structure

space (this is what the "K" in name Kuser indicates). As most embedded microcontroller applications do not use an operating system, we can place all our programs and data in the kernel address space and do not use the user address space at all. As a result of this, we will have 2 GB of address space for our programs and data.

The kernel address space is divided into four segments of 512 MB each: KSEG0, KSEG1, KSEG2, and KSEG3. The kernel address space contains the following memory areas:

• Program memory (flash)
• Data memory (RAM)
• Special function registers (SFR)
• Boot memory area

The kernel virtual address space contains two identical subsections, both translated to the same physical memory addresses. One of these subsections is *cacheable*, while the other one is not cacheable. Here, cacheable means that the instructions and data stored in the cache can be prefetched, and this speeds up the execution time considerably in sequential operations by eliminating the need to fetch data or instructions from the memory (the cache memory is a small, very fast memory). KSEG0 corresponds to the cacheable kernel address space, while KSEG1 corresponds to the noncacheable kernel address space. Each of KSEG0 and KSEG1 contains the program memory (flash), data memory (RAM), SFR, and boot memory area. Notice that a PIC32 microcontroller can run a program that is stored in the RAM memory (as opposed to the usual case of a program stored in the flash memory).

The cacheable kernel segment is used during normal program executions where instructions and data are fetched with the cache enabled. The noncacheable kernel segment is used during the processor initialisation routines where we wish to execute the instructions sequentially and with no cache present. The prefetch cache module can be enabled under software control.

When the PIC32 microcontroller is reset, it goes to the reset address of 0xBFC00000, which is the starting address of the boot program *in the noncacheable* kernel segment. The code in the boot location takes care of initialisation tasks, sets configuration bits, and then calls the user-written program in the cacheable kernel segment of the memory (the user program usually, but not always, starts from address 0x9D001000 in KSEG0 space). Notice that a new PIC32 microcontroller contains no program or data when it arrives from the factory. The boot program is normally placed there by the microcontroller programmer device.

The virtual or physical memory addresses are important to programmers writing assembly code, or developing DMA routines. In normal C programs, we are not concerned with the memory addresses as the compiler and linker take care of all the program and data placements.

2.1.2 The Microcontroller Clock

The block diagram of the clock circuit of the PIC32MX microcontroller is shown in Figure 2.5.

There are five clock sources: two of them use internal oscillators and three require external crystals or oscillator circuits. Three clock outputs are available: CPU system clock, USB clock, and peripheral clock.

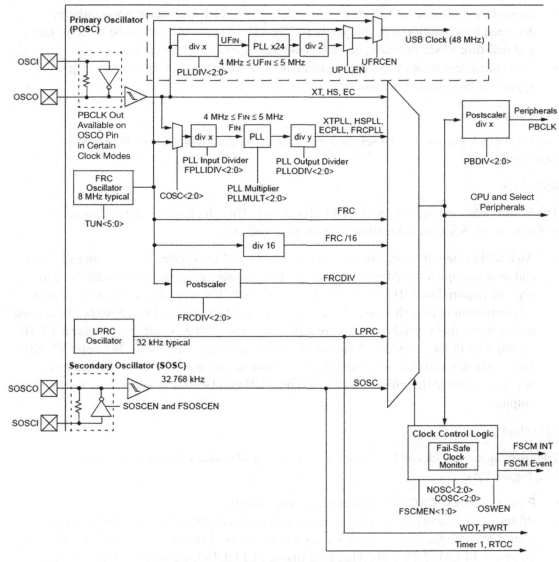

Figure 2.5: The PIC32MX Clock Block Diagram

Clock sources

- FRC is an internal oscillator requiring no external components. The clock frequency is 8 MHz and is designed for medium-speed, low-power operations. The clock can be tuned via the TUN bits of register OSCTUN. The clock is accurate to about ±2% after calibration.
- LPRC is a low-frequency, low-power internal oscillator with a frequency 32 kHz, requiring no external components.
- POSC is the primary high-speed oscillator, requiring an external crystal up to 20 MHz. The crystal is connected directly to the OSCI and OSCO inputs with two capacitors.
- Secondary oscillator (SOSC) is the secondary low-speed, high-accuracy oscillator, designed for operation with a crystal of 32,768 Hz. This oscillator can be used for timer and real-time clock modules, or as the main low-speed CPU oscillator.
- EC is the external clock source, requiring no external crystal. A square wave signal is applied to this input at the required frequency.

Clock outputs

The three clock outputs can be selected in various configurations as described below (see Figure 2.5).

USB clock

The USB operation requires an exact 48 MHz clock. This clock can be derived in many different ways. Some of the methods are given as follows:

- An 8 MHz external crystal–driven primary clock (POSC), divided by 2 (using PLLDIV), and then multiplied by 24 to give 96 MHz. This frequency can then be divided by 2 to give the required 48 MHz clock. UPLLEN and UFRCEN select this clock at the output. It is important to note that the input to PLLx24 must be between 4 and 5 MHz. We can use different external crystals with different PLL divisions. For example, we can use a 4 MHz crystal with POSC, divide by 1 (using PLLDIV), and then multiply by 24 to give 96 MHz. This frequency can then be divided by 2 as above to give the required 48 MHz clock.
- An external 48 MHz primary oscillator. UPLLEN and UFRCEN select this clock at the output.

CPU clock

The main system clock can be chosen from a variety of sources shown as follows (see Figure 2.5):

- Primary clock (POSC) selected at the output directly.
- Primary clock selected through the PLL input and output dividers, and the PLL multiplier. For example, if we use 8 MHz crystal, divide by 2 (using FPLLIDIV), multiply by 16 (using PLLMULT), and divide by 2 (using PLLODIV), we get a 32 MHz clock rate.

- Internal 8 MHz clock (FRC) can be selected directly at the output to give 8 MHz. Alternatively, the clock can be divided by 16, or it can pass through a postscaler to select the required frequency.
- Internal 8 MHz clock (FRC) can be selected with the PLL. For example, this clock can be divided by 2 to give 4 MHz at the input of the PLL (remember, the input clock to PLL must be between 4 and 5 MHz); it can be multiplied by 16, and then divided by 2 to give 32 MHz at the output.
- The 32 kHz clock (LPRC) can be selected directly at the output.
- The 32,768 Hz clock (SOSC) can be selected directly at the output.

Peripheral clock

The peripheral clock (PBCLK) is derived from the CPU system clock by passing it through a postscaler. The postscaler rate can be selected as 1, 2, 4, or 8. Thus, for example, if the CPU system clock is chosen as 80 MHz, then the peripheral clock can be 80, 40, 20, or 10 MHz.

Configuring the operating clocks

The clock configuration bits shown in Figure 2.5 can be selected by programming the SFR registers (e.g., OSCCON, OSCTUN, OSCCONCLR) or the device configuration registers (e.g., DEVCFG1 and DEVCFG2) during run time.

Alternatively, the operating clocks can be selected during the programming of the microcontroller chip. Most programming devices give options to users to select the operating clocks by modifying the device configuration registers just before the chip is programmed.

The crystal connections to OSCI and OSCO pins are shown in Figure 2.6 with two small capacitors. The connection of an external clock source in EC mode is shown in Figure 2.7. In this mode, the OSCO pin can be configured either as a clock output or as an I/O port.

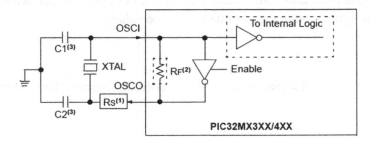

Note: 1. A series resistor, Rs, may be required for AT strip cut crystals.
2. The internal feedback resistor, RF, is typically in the range of 2–10 MΩ.
3. Refer to the *PIC32MX Family Reference Manual* (DS61132) for help in determining the best oscillator components.

Figure 2.6: Crystal Connection

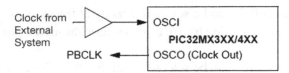

Figure 2.7: External Clock Connection

Performance and power consumption considerations

In microcontroller-based applications, the power consumption and performance are in direct conflict with each other. To lower the power consumption in an application, we also have to lower the performance, that is, the clock rate. Conversely, an increase in the performance also increases the power consumption. Thus, the higher the clock speed, the higher is the power consumption of the device. The designers of the PIC32 microcontroller have spent a considerable amount of time to provide a wide range of clock selection mechanisms so that the user can choose the best clock rate for the required power consumption.

For example, if an application can run at 8 MHz to do a job, there is no point in running the application at 80 MHz. In some applications, the device may be in standby mode and may be waiting for some user action (e.g., pressing a button). In such applications, a low clock rate can be selected while the device is in standby mode, and then a full high-speed clock rate can be selected when the device wakes up to run the actual application.

The flash wait states

The number of wait states is normally set by default to the highest value as this provides the safest operation. The SFR register CHECON, bits PFMWS, controls the number of wait states, and we can reduce its value for higher performance. It is important to remember, however, that setting wrong number of wait states could cause errors while accessing the flash memory. This register can take values between 0 and 7. For optimum performance, the wait states would be programmed to the minimum possible value.

2.1.3 Resets

There are several sources that can cause the microcontroller to reset. The following is a list of these sources:

- POR
- External reset (MCLR)
- Software reset (SWR)
- Watchdog timer reset (WDTR)
- BOR
- Configuration mismatch reset (CMR)

The SFR register RCON is the reset controller register, and the source of a reset can easily be found by reading the bits of this register. Any set bit in this register indicates that a reset has occurred, and depending on the position of this bit one can tell the actual source of the reset. For example, if bit 7 is set, then the source of reset is the external MCLR input (further details can be obtained from the individual microcontroller data sheets).

In this section, we are interested in external resets that are caused, for example, by the user pressing a button. The MCLR pin is used for external device reset, device programming, and device debugging. External reset occurs when the MCLR pin is lowered. Notice that the microcontroller can be reset in software by executing a specific sequence of operations (see the individual device data sheets).

2.1.4 The Input/Output Ports

The general-purpose I/O ports are very important in many applications as they allow the microcontroller to monitor and control devices attached to it. Although the I/O ports of the PIC32 microcontrollers have some similarities with the 8- and 16-bit devices, PIC32 microcontrollers offer greater functionality and new added features.

Some of the key features of PIC32 I/O ports are the following:

- Open-drain capability of each pin
- Pull-up resistors at each input pin
- Fast I/O bit manipulation
- Operation during CPU SLEEP and IDLE modes
- Input monitoring and interrupt generation on mismatch conditions

Figure 2.8 shows the basic block diagram of an I/O port.

An I/O port is controlled with the following SFR registers:

- TRISx: data direction control register for port x
- PORTx: port register for port x
- LATx: latch register for port x
- ODCx: open-drain control register for port x
- CNCON: interrupt-on-change control register

TRISx

There is a TRISx register for every port x. The TRISx register configures the port signal directions. A TRISx bit set to 1 configures the corresponding port pin "x" as an input. Similarly, a TRISx bit cleared to 0 configures the corresponding port pin "x" as an output. In the example shown in Figure 2.9, odd bits (1, 3, 5, 7, 9, 11, 13, and 15) of PORTB are configured as inputs and the remaining pins as outputs.

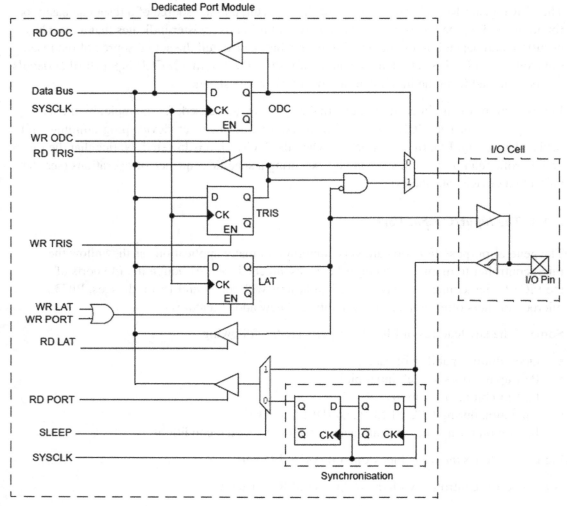

Figure 2.8: Block Diagram of an I/O Port

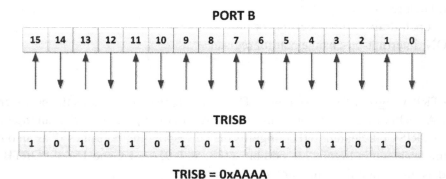

TRISB = 0xAAAA

Figure 2.9: Example TRIS Register Setting

PORTx

PORTx are the actual port registers. A write to PORTx sends data to the port latch register LATx, and this data appears at the output of the port. A read from a PORTx register reads the actual data at the output pin of the port (the state of the output pin may be affected by a device connected to the pin).

LATx

LATx are the port latch registers. A write to a latch register is same as sending data to the port register PORTx. A read from the latch register, however, reads the data present at the output latch of the port pin, and this may not be same as the actual state of the port pin. For example, the port pin may be pulled low by an external device. Under such circumstances, the latch register will read the data held at the output latch and not the actual state of the pin.

ODCx

Each I/O output pin can be configured individually as either normal output or open-drain output. SFR register ODCx controls the output state of a pin as follows: an OCDx bit set to 1 configures the corresponding port pin "x" as open-drain. Similarly, an OCDx bit cleared to 0 configures the corresponding port pin "x" as normal digital output. The open-drain feature allows the generation of outputs higher than the supply voltage (VDD) on any desired digital output pin. Notice that external pull-up resistors are normally used on open-drain output pins. In the example in Figure 2.10, all odd-numbered (1, 3, 5, 7, 9, 11, 13, and 15) PORTB output pins are configured to be open-drain.

CNCON

Some of the I/O pins can be configured to generate an interrupt when a change is detected on the pin. The SFR registers that control this change notice are the following:

- CNCON: used to enable or disable the interrupt-on-change (change notice) feature
- CNENx: contains the control bits, where "x" is the number of the change notice pin
- CNPUEx: enables or disables pull-up resistors on port pin "x"

Notice that bit CNIE (bit 0) of the IEC1 SFR register must be set to 1 to enable interrupt-on-change feature.

SET, CLR, INV I/O port registers

In addition to the I/O port registers described in this section, each port register has SET, CLR, and INV registers associated with it that can be useful in bit manipulation operations, allowing faster operations to be carried out.

ODCB = 0x5555

Figure 2.10: Example ODC Register Setting

The following registers are available for bit manipulation:

- TRISxSET
- TRISxCLR
- TRISxINV
- PORTxSET
- PORTxCLR
- PORTxINV
- LATxSET
- LATxCLR
- LATxINV
- ODCxSET
- ODCxCLR
- ODCxINV
- CNCONSET
- CNCONCLR
- CNCONINV
- CNPUESET
- CNPUECLR
- CNPUEINV

An example is given below to show how the SET, CLR, and INV registers can be used on TRISB register. Use of other registers is similar and is not repeated here.

Register TRISBCLR is a write-only register, and it clears selected bits in TRISB register. Writing a 1 in one or more bit positions clears the corresponding bits in TRISB register. Writing a 0 does not affect the register. For example, to clear bits 15, 5, and 0 of TRISB register, we issue the command: TRISBCLR = 0b1000000000100001 or TRSBCLR = 0x08021.

Register TRISBSET is a write-only register, and it sets selected bits in TRISB register. Writing a 1 in one or more bit positions sets the corresponding bits in TRISB register. Writing a 0 does not affect the register. For example, to set bits 15, 5, and 0 of TRISB register, we issue the command: TRISBSET = 0b1000000000100001 or TRSBSET = 0x08021.

Register TRISBINV is a write-only register, and it inverts selected bits in TRISB register. Writing a 1 in one or more bit positions inverts the corresponding bits in TRISB register. Writing a 0 does not affect the register. For example, to invert bits 15, 5, and 0 of TRISB register, we issue the command: TRISBINV = 0b1000000000100001 or TRSBINV = 0x08021.

Digital/analogue inputs

By default, all I/O pins are configured as analogue inputs. Setting the corresponding bits in the AD1PCFG register to 0 configures the pin as an analogue input, independent of the TRIS register setting for that pin. Similarly, setting a bit to 1 configures the pin as digital I/O. For example, to set all analogue pins as digital, we have to issue the command AD1PCFG = 0xFFFF.

It is recommended that any unused I/O pins that are not used should be set as outputs (e.g., by clearing the corresponding TRIS bits) and cleared to Low in software. When attaching devices to an I/O port, it is important to note with care that the normal I/O voltage of a pin is 3.6 V. An input pin can tolerate an input voltage up to 5 V, but the output voltage from an output pin cannot exceed 3.6 V.

2.1.5 The Parallel Master Port (PMP)

The PMP is an 8/16-bit parallel I/O port that can be used to communicate with a variety of parallel devices such as microcontrollers, external peripheral devices, LCDs, GLCDs, and so on. In this section, we will briefly look at the operation of the PMP. Further details about the PMP can be obtained from individual device data sheets.

Most microcontroller-based applications require address and data lines, and chip select control lines. The PMP provides up to 16 address and data lines, and up to 2 chip select lines. Addresses can be autoincremented and autodecremented for greater flexibility. In addition, parallel slave port support and individual read and write strobes are provided. Figure 2.11 shows a typical application of the PMP to interface to an external EPROM memory.

There are a number of SFR registers used to control the PMP. These are summarised as follows:

*   PMCON: PMP control register
*   PMMODE: PMP mode control register

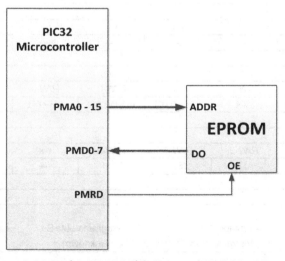

Figure 2.11: Using PMP With External EPROM Memory

- PMADDR: PMP address register
- PMDOUT: PMP data output register
- PMDIN: PMP data input register
- PMAEN: PMP address enable register
- PMSTAT: PMP status register

Each register has bit manipulation options. Thus, for example, in addition to PMCON, there are registers named PMCONCLR, PMCONSET, and PMCONINV.

PMCON

Register PMCON controls the PMP module. Figures 2.12 and 2.13 show the PMCON bit configuration and bit definitions, respectively. The bits in PMCON control address multiplexing, enable port control signals, enable chip select signals, and select signal polarity.

PMMODE

Register PMMODE controls the operational modes of the PMP. Figures 2.14 and 2.15 show the PMMODE bit configuration and bit definitions, respectively.

r-x	r-x	r-x	r-x	r-x	r-x	r-x	r-x
—	—	—	—	—	—	—	—
Bit 31							Bit 24

r-x	r-x	r-x	r-x	r-x	r-x	r-x	r-x
—	—	—	—	—	—	—	—
Bit 23							Bit 16

R/W-0	R/W-0	R/W-0	R/W-0	R/W-0	R/W-0	R/W-0	R/W-0
ON	FRZ	SIDL	ADRMUX1	ADRMUX0	PMPTTL	PTWREN	PTRDEN
Bit 15							Bit 8

R/W-0	R/W-0	R/W-0	R/W-0	R/W-0	r-x	R/W-0	R/W-0
CSF1	CSF0	ALP	CS2P	CS1P	—	WRSP	RDSP
Bit 7							Bit 0

Legend:			
R = Readable Bit	W = Writable Bit	P = Programmable Bit	r = Reserved Bit
U = Unimplemented Bit	-n = Bit Value at POR: ("0", "1", x = Unknown)		

Figure 2.12: PMCON Register Bit Configuration

PMADDR

Register PMADDR contains address of the external device and the chip select control bits. Figure 2.16 shows the PMADDR bit configuration and bit definitions.

PMDOUT

Register PMDOUT controls the buffered data output in slave mode. Figure 2.17 shows the bit configuration and bit definitions.

PMDIN

This register controls the I/O data ports in 8/16-bit master mode, and input data port in 8-bit slave mode. Figure 2.18 shows the bit configuration and bit definitions.

PMAEN

This register controls the operation of address and chip select pins of the PMP module. Figure 2.19 shows the bit configuration and bit definitions.

Bits 31–16	**Reserved:** Write '0'; Ignore Read
bit 15	**ON:** Parallel Master Port Enable bit
	1 = PMP enabled
	0 = PMP disabled, no off-chip access performed
	Note: When using 1:1 PBCLK divisor, the user's software should not read/write the peripheral's SFRs in the SYSCLK cycle immediately following the instruction that clears the module's ON bit.
bit 14	**FRZ:** Freeze in Debug Exception Mode bit
	1 = Freeze operation when CPU is in Debug Exception mode
	0 = Continue operation even when CPU is in Debug Exception mode
	Note: FRZ is writable in Debug Exception mode only; it is forced to '0' in normal mode.
bit 13	**SIDL:** Stop in IDLE Mode bit
	1 = Discontinue module operation when device enters IDLE mode
	0 = Continue module operation in IDLE mode
Bits 12–11	**ADRMUX<1:0>:** Address/Data Multiplexing Selection bits
	11 = All 16 bits of address are multiplexed on PMD<15:0> pins
	10 = All 16 bits of address are multiplexed on PMD<7:0> pins
	01 = Lower 8 bits of address are multiplexed on PMD<7:0> pins; upper 8 bits are on PMA<15:8>
	00 = Address and data appear on separate pins
bit 10	**PMPTTL**: PMP Module TTL Input Buffer Select bit
	1 = PMP module uses TTL input buffers
	0 = PMP module uses Schmidt Trigger input buffer
bit 9	**PTWREN:** Write Enable Strobe Port Enable bit
	1 = PMWR/PMENB port enabled
	0 = PMWR/PMENB port disabled
bit 8	**PTRDEN:** Read/Write Strobe Port Enable bit
	1 = PMRD/PMWR port enabled
	0 = PMRD/PMWR port disabled

Figure 2.13: PMCON Bit Definitions *(Continued)*

Register 13-1:   PMCON: Parallel Port Control Register (Continued)

Bits 7–6   **CSF<1:0>:** Chip Select Function Bits*

                  11 = Reserved
                  10 = PMCS2 and PMCS1 function as Chip Select
                  01 = PMCS2 functions as Chip Select; PMCS1 functions as address bit 14
                  00 = PMCS2 and PMCS1 function as address bits 15 and 14

bit 5   **ALP:** Address Latch Polarity bit[1]

                  1 = Active-high (PMALL and PMALH)
                  0 = Active-low ($\overline{\text{PMALL}}$ and $\overline{\text{PMALH}}$)

bit 4   **CS2P:** Chip Select 1 Polarity bit[1]

                  1 = Active-high (PMCS2)
                  0 = Active-low ($\overline{\text{PMCS2}}$)

bit 3   **CS1P:** Chip Select 0 Polarity bit[1]

                  1 = Active-high (PMCS1)
                  0 = Active-low ($\overline{\text{PMCS1}}$)

bit 2   **Reserved:** Write '0'; ignore read

bit 1   **WRSP:** Write Strobe Polarity bit
                For Slave Modes and Master mode 2 (PMMODE<9:8> = 00,01,10):
                  1 = Write strobe active-high (PMWR)
                  0 = Write strobe active-low ($\overline{\text{PMWR}}$)
                For Master mode 1 (PMMODE<9:8> = 11):
                  1 = Enable strobe active-high (PMENB)
                  0 = Enable strobe active-low (PMENB)

bit 0   **RDSP:** Read Strobe Polarity bit
                For Slave modes and Master mode 2 (PMMODE<9:8> = 00,01,10):
                  1 = Read Strobe active-high (PMRD)
                  0 = Read Strobe active-low ($\overline{\text{PMRD}}$)
                For Master mode 1 (PMMODE<9:8> = 11):
                  1 = Read/write strobe active-high (PMRD/$\overline{\text{PMWR}}$)
                  0 = Read/write strobe active-low ($\overline{\text{PMRD}}$/PMWR)

*These bits have no effect when their corresponding pins are used as address lines.

Figure 2.13: *(cont.)*

PMSTAT

This register contains the status bits when operating in buffered mode. Figures 2.20 and 2.21 show the bit configuration and bit definitions, respectively.

2.1.6 Timers

The PIC32 microcontroller supports five timers, named TIMER1–TIMER5. Timer 1 is 16-bit wide, while the other timers can be combined for 32-bit operation. In this section, we will look at the operation of Timer 1.

r-x	r-x	r-x	r-x	r-x	r-x	r-x	r-x
—	—	—	—	—	—	—	—
Bit 31							Bit 24

r-x	r-x	r-x	r-x	r-x	r-x	r-x	r-x
—	—	—	—	—	—	—	—
Bit 23							Bit 16

R-0	R/W-0	R/W-0	R/W-0	R/W-0	R/W-0	R/W-0	R/W-0
BUSY	IRQM<1:0>		INCM<1:0>		MODE16	MODE<1:0>	
Bit 15							Bit 8

R/W-0	R/W-0	R/W-0	R/W-0	R/W-0	R/W-0	R/W-0	R/W-0
WAITB<1:0>		WAITM<3:0>				WAITE<1:0>	
Bit 7							Bit 0

Legend:

R = Readable Bit	W = Writable Bit	P = Programmable Bit	r = Reserved Bit
U = Unimplemented Bit	-n = Bit Value at POR: ("0", "1", x = Unknown)		

Figure 2.14: PMMODE Register Bit Configuration

Bits 31–16   **Reserved:** Write '0'; Ignore Read

bit 15   **BUSY:** Busy bit (Master mode only)

     1 = Port is busy
     0 = Port is not busy

Bits 14–13   **IRQM<1:0>:** Interrupt Request Mode bits

     11 = Reserved; do not use
     10 = Interrupt generated when Read Buffer 3 is read or Write Buffer 3 is written (Buffered PSP mode) or on a read or write operation when PMA<1:0> =11 (Addressable Slave mode only)
     01 = Interrupt generated at the end of the read/write cycle
     00 = No Interrupt generated

Bits 12–11   **INCM<1:0>:** Increment Mode bits

     11 = Slave mode read and write buffers auto-increment (PMMODE<1:0> = 00 only)
     10 = Decrement ADDR<15:0> by 1 every read/write cycle*, †
     01 = Increment ADDR<15:0> by 1 every read/write cycle*, †
     00 = No increment or decrement of address

bit 10   **MODE16:** 8/16-bit Mode bit

     1 = 16-bit mode: a read or write to the data register invokes a single 16-bit transfer
     0 = 8-bit mode: a read or write to the data register invokes a single 8-bit transfer

Bits 9–8   **MODE<1:0>:** Parallel Port Mode Select bits

     11 = Master mode 1 (PMCSx, PMRD/PMWR, PMENB, PMA<x:0>, PMD<7:0>, and PMD<8:15>[‡])
     10 = Master mode 2 (PMCSx, PMRD, PMWR, PMA<x:0>, PMD<7:0>, and PMD<8:15>[‡])
     01 = Enhanced Slave mode, control signals (PMRD, PMWR, PMCS, PMD<7:0>, and PMA<1:0>)
     00 = Legacy Parallel Slave Port, control signals (PMRD, PMWR, PMCS, and PMD<7:0>)

Figure 2.15: PMMODE Bit Definitions *(Continued)*

Register 13-5:   PMMODE: Parallel Port Mode Register (Continued)

Bits 7–6   **WAITB1:WAITB0:** Data Setup to Read/Write Strobe Wait States bits[§]

11 =Data wait of 4 TPB; multiplexed address phase of 4 TPB
10 =Data wait of 3 TPB; multiplexed address phase of 3 TPB
01 =Data wait of 2 TPB; multiplexed address phase of 2 TPB
00 = Data wait of 1 TPB; multiplexed address phase of 1 TPB (DEFAULT)

Bits 5–2   **WAITM3:WAITM0:** Data Read/Write Strobe Wait States bits

1111 =Wait of 16 TPB
...
0001 =Wait of 2 TPB
0000 = Wait of 1 TPB (DEFAULT)

Bits 1–0   **WAITE1:WAITE0:** Data Hold After Read/Write Strobe Wait States bits[§]

11 =Wait of 4 TPB
10 =Wait of 3 TPB
01 =Wait of 2 TPB
00 =Wait of 1 TPB (DEFAULT)

for Read Operations:
11 =Wait of 3TPB
10 =Wait of 2TPB
01 =Wait of 1TPB
00 = Wait of 0TPB (DEFAULT)

[§] Whenever WAITM<3:0> = 0000, WAITB and WAITE bits are ignored and forced to 1 T_{PBCLK} cycle for a write operation; WAITB = 1 T_{PBCLK} cycle, WAITE = 0 T_{PBCLK} cycle for a read operation.

[*] Address bits A15 and A14 are not subject to auto-increment/decrement if configured as Chip Select CS2 and CS1.

[‡] These pins are active when bit MODE16 = 1 (16-bit mode).

[†] The PMPADDR register is always incremented/decremented by 1 regardless of the transfer data width.

Figure 2.15: *(cont.)*

Timer 1

Timer 1 is 16-bit wide that can be used in various internal and external timing and counting applications. This timer can be used in synchronous and asynchronous internal and external modes.

Figure 2.22 shows the block diagram of Timer 1. The source of clock for Timer 1 can be from the low-power SOSC, from the external input pin T1CK, or from the peripheral bus clock (PBCLK). A prescaler is provided with division ratios of 1, 8, 64, and 256 in order to change the timer clock frequency. The operation of the timer is given in the following paragraphs.

Timer register PR1 is loaded with a 16-bit number. Register TMR1 counts up at every clock pulse, and when PR1 is equal to TMR1, timer flag T1IF is set. At the same time, register TMR1 is reset to zero so that new count starts from zero again. If Timer 1 interrupts are enabled, then an interrupt will be generated where the program will jump to the interrupt service routine (ISR) whenever the timer flag is set.

r-x	r-x	r-x	r-x	r-x	r-x	r-x	r-x
—	—	—	—	—	—	—	—
Bit 31							Bit 24

r-x	r-x	r-x	r-x	r-x	r-x	r-x	r-x
—	—	—	—	—	—	—	—
Bit 23							Bit 16

R/W-0	R/W-0	R/W-0	R/W-0	R/W-0	R/W-0	R/W-0	R/W-0
CS2	CS1	ADDR<13:8>					
Bit 15							Bit 8

R/W-0	R/W-0	R/W-0	R/W-0	R/W-0	R/W-0	R/W-0	R/W-0
ADDR<7:0>							
Bit 7							Bit 0

Legend:

R = Readable Bit	W = Writable Bit	P = Programmable Bit	r = Reserved Bit
U = Unimplemented Bit	-n = Bit Value at POR: ("0", "1", x = Unknown)		

Bits 31–16   **Reserved:** Write '0'; ignore read

Bit 15   **CS2:** Chip Select 2 bit
                1 = Chip Select 2 is active
                0 = Chip Select 2 is inactive (pin functions as PMA<15>)

Bit 14   **CS1:** Chip Select 1 bit
                1 = Chip Select 1 is active
                0 = Chip Select 1 is inactive (pin functions as PMA<14>)

Bits 13–0   **ADDR<13:0>:** Destination Address bits

Figure 2.16: PMADDR Register Bit Configuration and Bit Definitions

Timer 1 is controlled by three registers: T1CON, PR1, and TMR1. Figures 2.23 and 2.24 show the bit configuration and bit definitions of register T1CON, respectively.

Assuming TMR1 starts counting from 0, the delay before the count reaches to PR1 is given by the following equation:

$$\text{Delay} = T \times \text{prescaler} \times \text{PR1}$$

where T is the clock period and PR1 is the value loaded into register PR1. Rearranging the above equation, we can find the value to be loaded into PR1 as:

$$\text{PR1} = \frac{\text{delay}}{T \times \text{prescaler}}$$

R/W-0	R/W-0	R/W-0	R/W-0	R/W-0	R/W-0	R/W-0	R/W-0
			DATAOUT<31:24>				
Bit 31							Bit 24

R/W-0	R/W-0	R/W-0	R/W-0	R/W-0	R/W-0	R/W-0	R/W-0
			DATAOUT<23:16>				
Bit 23							Bit 16

R/W-0	R/W-0	R/W-0	R/W-0	R/W-0	R/W-0	R/W-0	R/W-0
			DATAOUT<15:8>				
Bit 15							Bit 8

R/W-0	R/W-0	R/W-0	R/W-0	R/W-0	R/W-0	R/W-0	R/W-0
			DATAOUT<7:0>				
Bit 7							Bit 0

Legend:

R = Readable Bit	W = Writable Bit	P = Programmable Bit	r = Reserved Bit
U = Unimplemented Bit	-n = Bit Value at POR: ("0", "1", x = Unknown)		

Bits 31–0   **DATAOUT<31:0>:** Output Data Port bits for 8-bit write operations in Slave mode

Figure 2.17: PMDOUT Register Bit Configuration and Bit Definitions

R/W-0	R/W-0	R/W-0	R/W-0	R/W-0	R/W-0	R/W-0	R/W-0
			DATAIN<31:24>				
Bit 31							Bit 24

R/W-0	R/W-0	R/W-0	R/W-0	R/W-0	R/W-0	R/W-0	R/W-0
			DATAIN<23:16>				
Bit 23							Bit 16

R/W-0	R/W-0	R/W-0	R/W-0	R/W-0	R/W-0	R/W-0	R/W-0
			DATAIN<15:8>				
Bit 15							Bit 8

R/W-0	R/W-0	R/W-0	R/W-0	R/W-0	R/W-0	R/W-0	R/W-0
			DATAIN<7:0>				
Bit 7							Bit 0

Legend:

R = Readable Bit	W = Writable Bit	P = Programmable Bit	r = Reserved Bit
U = Unimplemented Bit	-n = Bit Value at POR: ("0", "1", x = Unknown)		

Bits 31–0   **DATAIN<31:0>:** Input and Output Data Port bits for 8 or 16-bit read/write operations in Master mode

Input Data Port for 8-bit read operations in Slave mode.

Figure 2.18: PMDIN Register Bit Configuration and Bit Definitions

r-x	r-x	r-x	r-x	r-x	r-x	r-x	r-x
—	—	—	—	—	—	—	—

Bit 31 — Bit 24

r-x	r-x	r-x	r-x	r-x	r-x	r-x	r-x
—	—	—	—	—	—	—	—

Bit 23 — Bit 16

R/W-0	R/W-0	R/W-0	R/W-0	R/W-0	R/W-0	R/W-0	R/W-0
			PTEN<15:8>				

Bit 15 — Bit 8

R/W-0	R/W-0	R/W-0	R/W-0	R/W-0	R/W-0	R/W-0	R/W-0
			PTEN<7:0>				

Bit 7 — Bit 0

Legend:
R = Readable Bit   W = Writable Bit   P = Programmable Bit   r = Reserved Bit
U = Unimplemented Bit   -n = Bit Value at POR: ("0", "1", x = Unknown)

Bits 31–16   **Reserved:** Write '0'; ignore read

Bits 15–14   **PTEN<15:14>: PMCSx Strobe Enable bits**
1 = PMA15 and PMA14 function as either PMA<15:14> or PMCS2 and PMCS1[1]
0 = PMA15 and PMA14 function as port I/O

Bits 13–2   **PTEN<13:2>: PMP Address Port Enable bits**
1 = PMA<13:2> function as PMP address lines
0 = PMA<13:2> function as port I/O

Bits 1–0   **PTEN<1:0>: PMALH/PMALL Strobe Enable bits**
1 = PMA1 and PMA0 function as either PMA<1:0> or PMALH and PMALL[2]
0 = PMA1 and PMA0 pads functions as port I/O

Figure 2.19: PMAEN Register Bit Configuration and Bit Definitions

An example is given here to illustrate the process. Assume that the clock frequency is 20 MHz, and it is required to generate a timer interrupt every 256 ms. Assuming a prescaler value of 256, the value to be loaded into PR1 is calculated as:

$$T = \frac{1}{20\,\text{MHz}} = 0.05 \times 10^{-3}\,\text{ms}$$

$$\text{PR1} = \frac{256\,\text{ms}}{0.05 \times 10^{-3} \times 256} = 20{,}000$$

Thus, the required settings are:

TMR1 = 0
Prescaler = 256
PR1 = 20,000 (0x4E20)

r-x	r-x	r-x	r-x	r-x	r-x	r-x	r-x
—	—	—	—	—	—	—	—
Bit 31							Bit 24

r-x	r-x	r-x	r-x	r-x	r-x	r-x	r-x
—	—	—	—	—	—	—	—
Bit 23							Bit 16

R-0	R/W-0	r-x	r-x	R-0	R-0	R-0	R-0
IBF	IBOV	—	—	IB3F	IB2F	IB1F	IB0F
Bit 15							Bit 8

R-1	R/W-0	r-x	r-x	R-1	R-1	R-1	R-1
OBE	OBUF	—	—	OB3E	OB2E	OB1E	OB0E
Bit 7							Bit 0

Legend:			
R = Readable Bit	W = Writable Bit	P = Programmable Bit	r = Reserved Bit
U = Unimplemented Bit	-n = Bit Value at POR: ("0", "1", x = Unknown)		

Figure 2.20: PMSTAT Register Bit Configuration

bits 31–16	**Reserved:** Write '0'; Ignore Read
bit 15	**IBF:** Input Buffer Full Status bit
	1 = All writable input buffer registers are full
	0 = Some or all of the writable input buffer registers are empty
bit 14	**IBOV:** Input Buffer Overflow Status bit
	1 = A write attempt to a full input byte buffer occurred (must be cleared in software)
	0 = No overflow occurred
	This bit is set (= 1) in hardware; can only be cleared (= 0) in software
bits 13–12	**Reserved:** Write '0'; ignore read
bit 11-8	**IBnF:** Input Buffer n Status Full bits
	1 = Input Buffer contains data that has not been read (reading buffer will clear this bit)
	0 = Input Buffer does not contain any unread data
bit 7	**OBE:** Output Buffer Empty Status bit
	1 = All readable output buffer registers are empty
	0 = Some or all of the readable output buffer registers are full
bit 6	**OBUF:** Output Buffer Underflow Status bit
	1 = A read occurred from an empty output byte buffer (must be cleared in software)
	0 = No underflow occurred
	This bit is set (= 1) in hardware; can only be cleared (= 0) in software
bits 5–4	**Reserved:** Write '0'; ignore read
bits 3–0	**OBnE:** Output Buffer n Status Empty bits
	1 = Output buffer is empty (writing data to the buffer will clear this bit)
	0 = Output buffer contains data that has not been transmitted

Figure 2.21: PMSTAT Register Bit Definitions

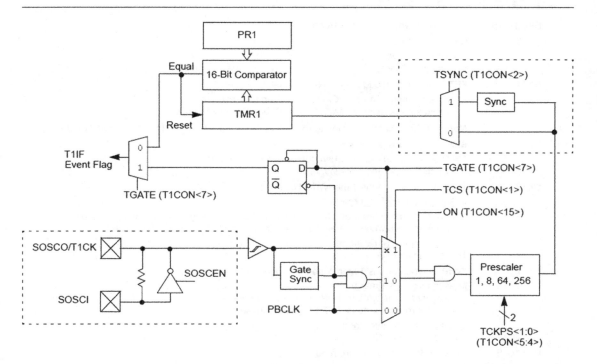

Note: The default State of the SOSCEN (OSCCON<1>) during a device Reset is controlled by the FSOSCEN bit in Configuration Word DEVCFG1.

Figure 2.22: Timer 1 Block Diagram

r-x	r-x	r-x	r-x	r-x	r-x	r-x	r-x
—	—	—	—	—	—	—	—
Bit 31							Bit 24

r-x	r-x	r-x	r-x	r-x	r-x	r-x	r-x
—	—	—	—	—	—	—	—
Bit 23							Bit 16

R/W-0	R/W-0	R/W-0	R/W-0	R-0	r-x	r-x	r-x
ON	FRZ	SIDL	TWDIS	TWIP	—	—	—
Bit 15							Bit 8

R/W-0	r-x	R/W-0	R/W-0	r-x	R/W-0	R/W-0	r-x
TGATE	—	TCKPS<1:0>		—	TSYNC	TCS	—
Bit 7							Bit 0

Figure 2.23: T1CON Bit Configuration

Bits 31–16	**Reserved:** Write '0'; Ignore Read
bit 15	**ON:** Timer On bit

1 = Timer is enabled
0 = Timer is disabled

bit 14 **FRZ:** Freeze in Debug Exception Mode bit

1 = Freeze operation when CPU is in Debug Exception mode
0 = Continue operation when CPU is in Debug Exception mode

Note: FRZ is writable in Debug Exception mode only; it is forced to "0" in normal mode

bit 13 **SIDL:** Stop in Idle Mode bit

1 = Discontinue operation when device enters Idle mode
0 = Continue operation in Idle mode

bit 12 **TWDIS:** Asynchronous Timer Write Disable bit
<u>In Asynchronous Timer mode:</u>

1 = Writes to asynchronous TMR1 are ignored until pending write operation completes
0 = Back-to-back writes are enabled (legacy asynchronous timer functionality)
<u>In Synchronous Timer mode:</u>
This bit has no effect

bit 11 **TWIP:** Asynchronous Timer Write in Progress bit
<u>In Asynchronous Timer mode:</u>
1 = Asynchronous write to TMR1 register in progress
0 = Asynchronous write to TMR1 register complete
<u>In Synchronous Timer mode:</u>
This bit is read as "0"

Bits 10–8 **Reserved:** Write "0"; ignore read

bit 7 **TGATE:** Gated Time Accumulation Enable bit
<u>When TCS = 1:</u>
This bit is ignored and read "0"
<u>When TCS = 0:</u>
1 = Gated time accumulation is enabled
0 = Gated time accumulation is disabled

bit 6 **Reserved:** Write "0"; ignore read

Bits 5–4 **TCKPS<1:0>:** Timer Input Clock prescaler Select bits

11 = 1:256 prescale value
10 = 1:64 prescale value
01 = 1:8 prescale value
00 = 1:1 prescale value

bit 3 **Reserved:** Write "0"; ignore read

bit 2 **TSYNC:** Timer External Clock Input Synchronization Selection bit
<u>When TCS = 1:</u>
1 = External clock input is synchronised
0 = External clock input is not synchronised
<u>When TCS = 0:</u>
This bit is ignored and read "0"

bit 1 **TCS:** Timer Clock Source Select bit

1 = External clock from T1CKI pin
0 = Internal peripheral clock

bit 0 **Reserved:** Write "0"; ignore read

Figure 2.24: T1CON Bit Definitions

The steps to configure Timer 1 are summarised as follows (assuming no timer interrupts are required):

- Disable Timer 1 (T1CON, bit 15 = 0).
- Select required prescaler value (T1CON, bits 4–5).
- Select timer clock (T1CON, bit 1).
- Load PR1 register with required value.
- Clear TMR1 register to 0.
- Enable the timer (T1CON, bit 15 = 1).

An example project is given in the Projects section of this book to show how to configure Timer 1 with interrupts.

Timers 2, 3, 4, 5

Although Timers 2, 3, 4, and 5 can operate as 16-bit timers, Timers 2 and 3, and Timers 4 and 5 can be combined to provide two 32-bit wide internal and external timers.

Figure 2.25 shows the block diagram of these timers in 16-bit mode (note that "x" represents the timer number and is between 2 and 5). Notice that, compared with Timer 1, the prescaler

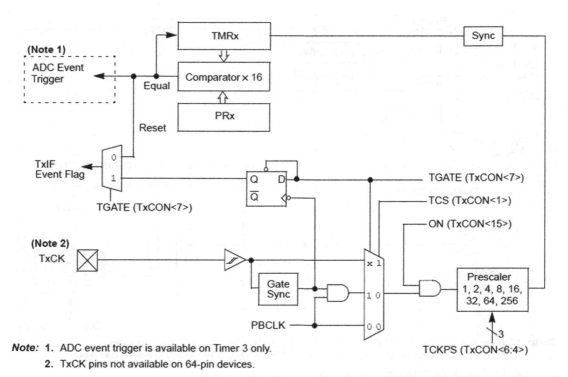

Note: 1. ADC event trigger is available on Timer 3 only.
2. TxCK pins not available on 64-pin devices.

Figure 2.25: Timers 2, 3, 4, and 5 in 16-Bit Mode

has been extended and the low-power SOSC has been removed. The operation of these timers in 16-bit mode is same as that of Timer 1.

Figure 2.26 shows the timers in 32-bit mode where Timers 2/3 and 4/5 are combined (in this figure "x" represents Timers 2–5 in 16-bit mode, while in 32-bit mode "x" represents Timer 2 or 4, while "y" represents Timer 3 or 5).

Considering, for example, Timers 2 and 3 in 32-bit operation, register PR3 and PR2 are combined to form a 32-bit register. Similarly, registers TMR3 and TMR2 are combined to form a 32-bit register. A 32-bit comparator is used to compare the two pairs and generate the T3IF flag when they are equal. In 32-bit mode, Timer 2/3 pair is controlled with register T2CON. Similarly, Timer 4/5 pair is controlled with register T4CON.

The steps to configure Timer 2/3 are summarised as follows (assuming no timer interrupts are required):

- Disable Timer 1 (T2CON, bit 15 = 0).
- Set 32-bit mode (T2CON, bit 3 = 1).
- Select required prescaler value (T2CON, bits 4–6).

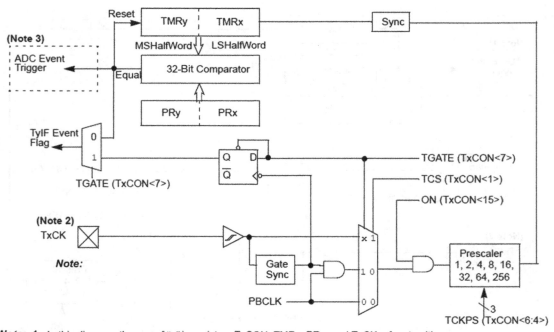

Note: 1. In this diagram, the use of "x" in registers TxCON, TMRx, PRx, and TxCK refers to either Timer 2 or 4; the use of "y" in registers TyCON, TMRy, PRy, and TyIF refers to either Timer 3 or 5.

2. TxCK pins not available on 64-pin devices.

3. ADC event trigger is available only on Timer 2/3 pair.

Figure 2.26: Timers 2/3 and 4/5 Combined for 32-Bit Operation

- Select internal timer clock (T2CON, bit 1 = 0).
- Clear timer registers TMR2 and TMR3.
- Load PR2 and PR3 registers with required value.
- Clear TMR1 register to 0.
- Enable the timer (T2CON, bit 15 = 1).

2.1.7 Real-Time Clock and Calendar

The RTCC module is intended for accurate real-time date and time applications. Some of the features of this module are:

- Provides real-time hours, minutes, seconds, weekday, date, month, and year
- Provides alarm intervals for 0.5, 1, 10 s, 1, 10 min, 1 h, 1 day, 1 week, 1 month, and 1 year
- Provides leap year correction
- Provides long-term battery operation
- Provides alarm pulse on output pin
- Requires an external 32,768 Hz crystal

Figure 2.27 shows a block diagram of the RTCC module. The module is controlled by the following six SFR registers:

- RTCCON
- RTCALRM

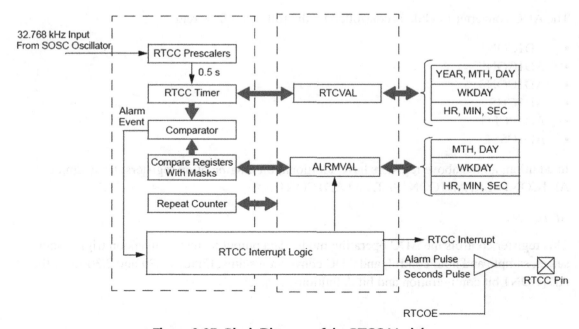

Figure 2.27: Block Diagram of the RTCC Module

- RTCTIME
- RTCDATE
- ALRMTIME
- ALRMDATE

Interested readers can find the programming details of the RTCC module in the individual microcontroller data sheets.

2.1.8 Analogue-to-Digital Converter

The PIC32MX460F512L microcontroller contains multiplexed 16-channel, 10-bit A/D converters. These converters have the following features:

- 500 ksps (kilo samples per second) conversion speed
- Multiplexed 16 channels (2 switchable multiplexers to select different analogue channels and different reference sources)
- Sample-and-hold amplifier (SHA)
- Automatic input channel scanning
- 32-Bit wide, 16-word result buffer
- Various conversion result formats (integer, signed, unsigned, 16- or 32-bit output)
- External voltage references
- Operation during CPU SLEEP and IDLE modes

Figure 2.28 shows a block diagram of the ADC converter module.

The ADC converter module is controlled by the following SFR registers:

- AD1CON1
- AD1CON2
- AD1CON3
- AD1CHS
- AD1PCFG
- AD1CSSL

In addition, all the above registers have additional bit manipulation registers, for example, AD1CON1CLR, AD1CON1SET, and AD1CON1INV.

AD1CON1

This register controls the ADC operating mode, data output format, conversion trigger source select, sample and hold control, and ADC conversion status. Figures 2.29 and 2.30 show the AD1CON1 bit configuration and bit definitions, respectively.

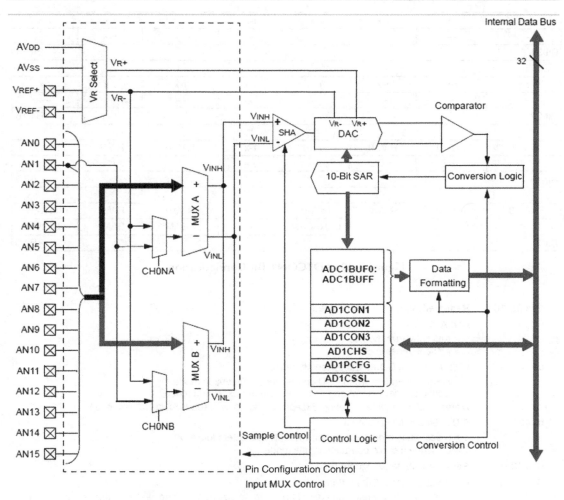

Figure 2.28: ADC Converter Module

AD1CON2

This register controls the voltage reference selection, scan input selection, ADC interrupt selection bits, ADC result buffer configuration, and alternate input sample mode selection. Figures 2.31 and 2.32 show the AD1CON2 bit configuration and bit definitions, respectively.

AD1CON3

This register controls the ADC clock selection. Figure 2.33 shows the bit configuration and bit definitions.

r-x	r-x	r-x	r-x	r-x	r-x	r-x	r-x
—	—	—	—	—	—	—	—
Bit 31							Bit 24

r-x	r-x	r-x	r-x	r-x	r-x	r-x	r-x
—	—	—	—	—	—	—	—
Bit 23							Bit 16

R/W-0	R/W-0	R/W-0	r-x	r-x	R/W-0	R/W-0	R/W-0
ON	FRZ	SIDL	—	—	FORM<2:0>		
Bit 15							Bit 8

R/W-0	R/W-0	R/W-0	R/W-0	r-x	R/W-0	R/W-0	R/C-0
SSRC<2:0>			CLRASAM	—	ASAM	SAMP	DONE
Bit 7							Bit 0

Figure 2.29: AD1CON1 Bit Configuration

Bits 31–16 **Reserved:** Write "0"; Ignore Read

bit 15 **ON:** ADC Operating Mode bit
1 = A/D converter module is operating
0 = A/D converter is off

bit 14 **FRZ:** Freeze in Debug Exception Mode bit
1 = Freeze operation when CPU enters Debug Exception mode
0 = Continue operation when CPU enters Debug Exception mode
Note: FRZ is writable in Debug Exception mode only. It reads "0" in Normal mode

bit 13 **SIDL:** Stop in Idle Mode bit
1 = Discontinue module operation when device enters Idle mode
0 = Continue module operation in Idle mode

Bits 12–11 **Reserved:** Write "0"; ignore read

Bits 10–8 **FORM<2:0>:** Data Output Format bits
011 = Signed Fractional 16-bit (DOUT = 0000 0000 0000 0000 sddd dddd dd00 0000)
010 = Fractional 16-bit (DOUT = 0000 0000 0000 0000 dddd dddd dd00 0000)
001 = Signed Integer 16-bit (DOUT = 0000 0000 0000 0000 ssss sssd dddd dddd)
000 = Integer 16-bit (DOUT = 0000 0000 0000 0000 0000 00dd dddd dddd)
111 = Signed Fractional 32-bit (DOUT = sddd dddd dd00 0000 0000 0000 0000 0000)
110 = Fractional 32-bit (DOUT = dddd dddd dd00 0000 0000 0000 0000 0000)
101 = Signed Integer 32-bit (DOUT = ssss ssss ssss ssss ssss sssd dddd dddd)
100 = Integer 32-bit (DOUT = 0000 0000 0000 0000 0000 00dd dddd dddd)

Bits 7–5 **SSRC<2:0>:** Conversion Trigger Source Select bits
111 = Internal counter ends sampling and starts conversion (auto convert)
110 = Reserved
101 = Reserved
100 = Reserved
011 = Reserved
010 = Timer 3 period match ends sampling and starts conversion
001 = Active transition on INT0 pin ends sampling and starts conversion
000 = Clearing SAMP bit ends sampling and starts conversion

Figure 2.30: AD1CON1 Bit Definitions *(Continued)*

bit 4   **CLRASAM:** Stop Conversion Sequence Bit (when the first A/D converter interrupt is generated)
1 = Stop conversions when the first ADC interrupt is generated. Hardware clears the ASAM bit when the ADC interrupt is generated
0 = Normal operation, buffer contents will be overwritten by the next conversion sequence

bit 3   **Reserved:** Write '0'; ignore read

bit 2   **ASAM:** ADC Sample Auto-Start bit
1 = Sampling begins immediately after last conversion completes; SAMP bit is automatically set
0 = Sampling begins when SAMP bit is set

bit 1   **SAMP:** ADC Sample Enable bit
1 = The ADC SHA is sampling
0 = The ADC sample/hold amplifier is holding
When ASAM = 0, writing '1' to this bit starts sampling
When SSRC = 000, writing '0' to this bit will end sampling and start conversion

bit 0   **DONE:** A/D Conversion Status bit
1 = A/D conversion is done
0 = A/D conversion is not done or has not started
Clearing this bit will not affect any operation in progress
Note: The DONE bit is not persistent in automatic modes. It is cleared by hardware at the beginning of the next sample.

Figure 2.30: *(cont.)*

r-x	r-x	r-x	r-x	r-x	r-x	r-x	r-x
—	—	—	—	—	—	—	—
Bit 31							Bit 24

r-x	r-x	r-x	r-x	r-x	r-x	r-x	r-x
—	—	—	—	—	—	—	—
Bit 23							Bit 16

R/W-0	R/W-0	R/W-0	R/W-0	r-x	R/W-0	r-x	r-x
VCFG<2:0>			OFFCAL	—	CSCNA	—	—
Bit 15							Bit 8

R-0	r-x	R/W-0	R/W-0	R/W-0	R/W-0	R/W-0	R/W-0
BUFS	—	SMPI<3:0>				BUFM	ALTS
Bit 7							Bit 0

Figure 2.31: AD1CON2 Bit Configuration

AD1CHS

This register selects the input channels for multiplexers A and B. Figures 2.34 and 2.35 show the bit configuration and bit definitions, respectively.

AD1PCFG

This register selects the input ports as analogue or digital. Setting a bit to 1 makes the corresponding ANx port pin a digital port. Similarly, clearing a bit to 0 makes the corresponding ANx port pin an analogue port. To configure all analogue ports as digital,

| Bits 31–16 | **Reserved:** Write '0'; Ignore Read |
| Bits 15–13 | **VCFG<2:0>:** Voltage Reference Configuration bits |

	ADC V$_{R+}$	ADC V$_{R-}$
000	AV$_{DD}$	AV$_{SS}$
001	External V$_{REF+}$ pin	AV$_{SS}$
010	AV$_{DD}$	External V$_{REF-}$ pin
011	External V$_{REF+}$ pin	External V$_{REF-}$ pin
1xx	AV$_{DD}$	AV$_{SS}$

bit 12
> **OFFCAL:** Input Offset Calibration Mode Select bit
>
> 1 = Enable Offset Calibration mode
> V$_{INH}$ and V$_{INL}$ of the SHA are connected to V$_{R-}$
> 0 = Disable Offset Calibration mode
> The inputs to the SHA are controlled by AD1CHS or AD1CSSL

bit 11
> **Reserved:** Write '0'; ignore read

bit 10
> **CSCNA:** Scan Input Selections for CH0+ SHA Input for MUX A Input Multiplexer Setting bit
> 1 = Scan inputs
> 0 = Do not scan inputs

Bits 9–8
> **Reserved:** Write '0'; ignore read

bit 7
> **BUFS:** Buffer Fill Status bit
> Only valid when BUFM = 1 (ADRES split into 2 – 8-word buffers)
> 1 = ADC is currently filling buffer 0x8-0xF; user should access data in 0x0-0x7
> 0 = ADC is currently filling buffer 0x0-0x7; user should access data in 0x8-0xF

bit 6
> **Reserved:** Write '0'; ignore read

Bits 5–2
> **SMPI<3:0>:** Sample/Convert Sequences Per Interrupt Selection bits
> 1111 = Interrupts at the completion of conversion for each 16th sample/convert sequence
> 1110 = Interrupts at the completion of conversion for each 15th sample/convert sequence
> . . . . .
> 0001 = Interrupts at the completion of conversion for each 2nd sample/convert sequence
> 0000 = Interrupts at the completion of conversion for each sample/convert sequence

bit 1
> **BUFM:** ADC Result Buffer Mode Select bit
> 1 = Buffer configured as two 8-word buffers, ADC1BUF(7...0), ADC1BUF(15...8)
> 0 = Buffer configured as one 16-word buffer ADC1BUF(15...0)

bit 0
> **ALTS:** Alternate Input Sample Mode Select bit
> 1 = Uses MUX A input multiplexer settings for first sample, and then alternates between MUX B
> and MUX A input multiplexer settings for all subsequent samples
> 0 = Always use MUX A input multiplexer settings

Figure 2.32: AD1CON2 Bit Definitions

issue the command AD1PCFG = 0xFFFF. Similarly, to configure ports as analogue, issue the command AD1PCFG = 0.

AD1CSSL

This register controls the ADC input scanning. Setting a bit of the register to 1 selects the corresponding ANx port pin for input scan. Similarly, clearing a bit of the register to 0 does not select the corresponding ANx port pin for input scan.

R/W-0	r-x	r-x	R/W-0	R/W-0	R/W-0	R/W-0	R/W-0
ADRC	—	—			SAMC<4:0>		
Bit 15							Bit 8

R/W-0	R/W-0	R/W-0	R/W-0	R/W-0	R/W-0	R/W-0	R/W-0
			ADCS<7:0>				
Bit 7							Bit 0

Legend:
R = Readable Bit   W = Writable Bit   P = Programmable Bit   r = Reserved Bit
U = Unimplemented Bit   -n = Bit Value at POR: ("0", "1", x = Unknown)

Bits 31–16   **Reserved:** Write "0"; ignore read

ADRC: ADC Conversion Clock Source bit
1 = ADC internal RC clock
0 = Clock derived from Peripheral Bus Clock (PBClock)

Bits 14–13   **Reserved:** Write "0"; ignore read

Bits 12–8   **SAMC<4:0>:** Auto Sample Time bits
11111 = 31 T_{AD}
. . . . .
00001 = 1 T_{AD}
00000 = 0 T_{AD} (Not allowed)

Bits 7–0   **ADCS<7:0>:** ADC Conversion Clock Select bits
11111111 = T_{PB} · (ADCS<7:0> + 1) · 2 = 512 · T_{PB} = T_{AD}
. . . . . .
00000001 = T_{PB} · (ADCS<7:0> + 1) · 2 = 4 · T_{PB} = T_{AD}
00000000 = T_{PB} · (ADCS<7:0> + 1) · 2 = 2 · T_{PB} = T_{AD}

Figure 2.33: AD1CON3 Bit Configuration and Bit Definitions

R/W-0	r-x	r-x	r-x	R/W-0	R/W-0	R/W-0	R/W-0
CH0NB	—	—	—		CH0SB<3:0>		
Bit 31							Bit 24

R/W-0	r-x	r-x	r-x	R/W-0	R/W-0	R/W-0	R/W-0
CH0NA	—	—	—		CH0SA<3:0>		
Bit 23							Bit 16

r-x	r-x	r-x	r-x	r-x	r-x	r-x	r-x
—	—	—	—	—	—	—	—
Bit 15							Bit 8

r-x	r-x	r-x	r-x	r-x	r-x	r-x	r-x
—	—	—	—	—	—	—	—
Bit 7							Bit 0

Figure 2.34: AD1CHS Bit Configuration

Bit 31   **CH0NB:** Negative Input Select for MUX B bit
                    1 = Channel 0 negative input is AN1
                    0 = Channel 0 negative input is VR–

Bits 30–28   **Reserved:** Write '0'; ignore read

Bits 27–24   **CH0SB<3:0>:** Positive Input Select for MUX B bits
                    1111 = Channel 0 positive input is AN15
                    1110 = Channel 0 positive input is AN14
                    1101 = Channel 0 positive input is AN13
                    . . . . . .
                    0001 = Channel 0 positive input is AN1
                    0000 = Channel 0 positive input is AN0

Bit 23   **CH0NA:** Negative Input Select for MUX A Multiplexer Setting bit[2]
                    1 = Channel 0 negative input is AN1
                    0 = Channel 0 negative input is VR–

Bits 22–20   **Reserved:** Write '0'; ignore read

Bits 19–16   **CH0SA<3:0>:** Positive Input Select for MUX A Multiplexer Setting bits
                    1111 = Channel 0 positive input is AN15
                    1110 = Channel 0 positive input is AN14
                    1101 = Channel 0 positive input is AN13
                    . . . . . .
                    0001 = Channel 0 positive input is AN1
                    0000 = Channel 0 positive input is AN0

Bits 15–0   **Reserved:** Write '0'; ignore read

Figure 2.35: AD1CHS Bit Definitions

Operation of the ADC module

The total ADC conversion time consists of the *acquisition time* and the *ADC conversion time*. The ADC has a single SHA. During the acquisition time, the analogue signal is sampled and held by the SHA. After the input is stable, it is disconnected from the SHA and the ADC conversion starts, which converts the analogue signal into digital. The conversion time is the actual time it takes the ADC converter to convert the signal held by the SHA. Although there is only one SHA, there are two multiplexers called MUXA and MUXB, controlled by register AD1CHS. The ADC converter can switch between MUXA and MUXB inputs.

The converted data in the result register can be read in eight different formats, controlled by AD1CON1.

Sampling can be started manually or automatically. In manual mode, bit 2 of AD1CON1 is cleared. Data acquisition is started when bit 1 of AD1CON1 is set. This bit must be set to re-start acquisition. In automatic mode, bit 2 of AD1CON is set, and acquisition starts automatically after a previous sample has completed.

The scan mode enables a number of input channels to be scanned and converted into digital. This mode is enabled by setting bit 10 of AD1CON2. Each bit in the AD1CSSL register corresponds to an analogue input channel, and if a bit is set to 1 in AD1CSSL, then the corresponding input channel is in the scan sequence.

The following steps summarise how to configure the ADC module, assuming no interrupts are to be generated (see the individual microcontroller data sheets for more information):

- Configure port pin as analogue by clearing the appropriate bits of AD1PCFG.
- Select the required analogue input channels using AD1CHS.
- Select the format of the result data using bits 8–10 of AD1CON1.
- Select the sample clock source using bits 5–7 of AD1CON1.
- Select the voltage reference source using bits 13–15 of AD1CON2.
- Select the scan mode (if required), alternating sample mode (if required), and autoconvert sample time (if required).
- Select the result buffer mode using bit 1 of AD1CON2.
- Select the A/D clock source using bit 15 of AD1CON3.
- Select A/D clock prescaler using bits 0–7 of AD1CON3.
- Turn ON the A/D using bit 15 of AD1CON1.

Most high-level compilers provide a built-in library for simple one-channel ADC conversions, where all the required registers are set automatically.

2.1.9 Interrupts

Interrupt control is one of the most complex parts of the PIC32 microcontrollers. There are 96 interrupt sources with up to 64 interrupt vectors, and a large number of interrupt control registers. The full description of the interrupt control is beyond the scope of this book, and interested readers should consult the individual microcontroller data sheets for much more information.

In this section, we shall be looking at the basic operation of the interrupt module and see how an interrupt-based program can be written using the mikroC PRO for PIC32 language.

The basic features of the PIC32 interrupt controller module are the following:

- Up to 96 interrupt sources
- Up to 64 interrupt vectors
- Single and multiple vector interrupt modes
- Seven user-selectable priority levels for each interrupt
- Four user-selectable subpriority levels within each priority
- User-configurable interrupt vector table location and spacing

The interrupt control module has the following SFR registers:

- INTCON
- INSTAT
- IPTMR
- IFS0, IFS1

- IEC0, IEC1
- IPC0–IPC11

In addition, all the above registers have additional bit manipulation registers, for example, INTCONCLR, INTCONSET, and INTCONINV. Register INTCON controls the interrupt vector mode and external interrupt edge mode. Other registers control the individual interrupt sources, such as enabling and disabling them.

PIC32 microcontrollers support both single and multiple vectored interrupts. In single vectored interrupts, all interrupting devices have the same common ISR addresses. The source of the interrupt is then determined by examining the interrupt flags of each interrupting source. This is actually the commonly used method in 8-bit microcontrollers. In multiple vectored interrupt operations, each interrupting device has its unique ISR address. (In actual fact, PIC32 microcontrollers have 96 interrupt sources and only 64 vectors. As such, some of the interrupts share the same vector.)

PIC32 microcontrollers support seven levels of interrupt priority (ipl1–ipl7). If more than one interrupt occurs at the same time, then the one with higher priority is serviced first. If while servicing an interrupt, a lower-priority interrupt occurs, then it will be ignored by the processor. In addition to standard interrupt priority levels, the PIC32 microcontrollers support four levels of subpriorities. Thus, should two interrupts at the same priority level interrupt at the same time, the one with the higher subpriority level will be serviced first. At reset or power-up, all interrupts are disabled and set to priority level ipl0.

In addition to general interrupt configuration bits, each interrupt source has associated control bits in the SFR registers, and these bits must be configured correctly for an interrupt to be accepted by the CPU. Some of the important interrupt source control bits are the following:

- Each interrupting source has an *interrupt enable bit* (denoted by suffix -IE) in the device data sheet. This bit must be set to 1 for an interrupt to be accepted from this source (at reset or power-up, all interrupt enable bits are cleared). Some commonly used interrupt enable bits are given in the following table:

Interrupt Source	Interrupt Enable Bit	Bit Position	Register
Timer 1	T1IE	4	IEC0
External Int 0	INT0IE	3	IEC0
External Int 1	INT1IE	7	IEC0
External Int 2	INT2IE	11	IEC0
External Int 3	INT3IE	15	IEC0

- Each interrupting source has an *interrupt flag* (denoted by -IF) in the device data sheet. This bit is set automatically when an interrupt occurs, and must be cleared in software before any more interrupts can be accepted from the same source. The bit is usually cleared inside the ISR of the interrupting source. The interrupt flags of some commonly used interrupt sources are given as follows:

Interrupt Source	Interrupt Flag Bit	Bit Position	Register
Timer 1	T1IF	4	IFS0
External Int 0	INT0IF	3	IFS0
External Int 1	INT1IF	7	IFS0
External Int 2	INT2IF	11	IFS0
External Int 3	INT3IF	15	IFS0

- Each interrupting source has a priority level from 1 to 7 (priority level 0 disabled the interrupt), and subpriority levels from 0 to 3. The priority levels are denoted by suffix -IP, and the subpriority levels by suffix -IS. The priority levels of some commonly used interrupt sources are given as follows:

Interrupt Source	Interrupt Priority Level	Bit Positions	Register
Timer 1	T1IP	2–4	IPC1
External Int 0	INT0IP	26–28	IPC0
External Int 1	INT1IP	26–28	IPC1
External Int 2	INT2IP	26–28	IPC2
External Int 3	INT3IP	26–28	IPC3

Interrupt Source	Interrupt Subpriority Level	Bit Positions	Register
Timer 1	T1IS	0–1	IPC1
External Int 0	INT0IS	24–25	IPC0
External Int 1	INT1IS	24–25	IPC1
External Int 2	INT2IS	24–25	IPC2
External Int 3	INT3IS	24–25	IPC3

Some examples are given in the following to show how the timer interrupts and external interrupts can be configured. The Projects section of this book gives real examples on configuring and using both timer and external interrupts.

Configuring Timer 1 interrupts

Timer 1 counts up until the value in TMR1 matches the one in period register PR1, and then the interrupt flag T1IF is set automatically. If the Timer 1 interrupt enable flag T1IE is set, then an interrupt will be generated to the processor. In multi-interrupt operations, it is recommended to set the priority and subpriority levels of the timer interrupt.

The steps in configuring the Timer 1 interrupts are given as follows:

- Configure for single vector or multivector interrupt mode (by default, after reset or power-up the single vector mode is selected).

- Disable Timer 1, ON (bit 15 in T1CON).
- Clear timer register TMR1 to 0.
- Select timer prescaler (bits 4–5 in T1CON).
- Select timer clock source, TCS (bit 1 in T1CON).
- Load period register PR1 as required.
- Clear interrupt flag, T1IF (bit 4 in IFS0).
- Set Timer 1 priority level, T1IP (bits 2–4 in IPC1).
- Set Timer 1 subpriority level, T1IS (if required).
- Enable Timer 1 interrupts, T1IE (bit 4 in IEC0).
- Enable Timer 1, ON (bit 15 in T1CON).
- Write the ISR.
- Enable interrupts.

Configuring external interrupt 0

The PIC32 microcontroller supports four external interrupts inputs, INT0–INT3. Interrupts can be recognised on either the low-to-high or the high-to-low transition of the interrupt pin, selected by INTCON. External interrupt flag INT0IF must be cleared before an interrupt can be accepted. In addition, INT0IE bit must be set to enable interrupts from external interrupt INT0 pin. In multi-interrupt operations, it is recommended to set the priority and subpriority levels of the interrupt.

The steps in configuring the external interrupt 0 are given as follows:

- Configure for single vector or multivector interrupt mode (bit 12, INTCON). By default, after reset or power-up the single vector mode is selected.
- Clear external interrupt 0 flag, INT0IF (bit 3 in IFS0).
- Set required interrupt edge, INT0EP (bit 0 in INTCON).
- Set external interrupt 0 priority level, INT0IP (bits 26–28 in IPC0).
- Set external interrupt subpriority level, INT0IS (if required).
- Enable external interrupt 0, INT0IE (bit 3 in IEC0).
- Write the ISR.
- Enable interrupts.

PIC32 interrupt service routines

The PIC32 microcontroller interrupt controller can be configured to operate in one of the following two modes:

- *Single vector mode*: All interrupt requests will be serviced at one vector address (mode out of reset).
- *Multivector mode*: Interrupt requests will be serviced at the calculated vector address.

In single vector mode, the CPU always vectors to the same address. This means that only one ISR can be defined. The single vector mode address is calculated by using the exception base (EBase) address (its address default is 0x9FC01000E). The exact formula for single vector mode is as follows (see the PIC32 microcontroller individual data sheets for more information):

Single vector address = EBase + 0x200

In multivector mode, the CPU vectors to the unique address for each vector number. Each vector is located at a specific offset, with respect to a base address specified by the EBase register in the CPU. The individual vector address offset is determined by the following equation:

EBase + (Vector _ Number × Vector _ Space) + 0x200

The PIC32 family of devices employs two register sets: a *primary register set* for normal program execution and a *shadow register set* for highest-priority interrupt processing.

- *Register set selection in single vector mode*: In single vector mode, you can select which register set will be used. By default, the interrupt controller will instruct the CPU to use the first register set. This can be changed later in the code.
- *Register set selection in multivector mode*: When a priority level interrupt matches a shadow set priority, the interrupt controller instructs the CPU to use the shadow set. For all other interrupt priorities, the interrupt controller instructs the CPU to use the primary register set.

In order to correctly utilise interrupts and correctly write the ISR code, the user will need to take care of the following:

1. Write the ISR.
2. Initialise the module that will generate an interrupt.
3. Set the correct priority and subpriority for the used module according to the priorities set in the ISR.
4. Enable interrupts.

2.2 Summary

This chapter has described the architecture of the PIC32 family of microcontrollers. PIC32MX360F512L was taken as a typical example microcontroller in the family.

Various important parts and peripheral circuits of the PIC32 series of microcontrollers have been described, including the data memory, program memory, clock circuits, reset circuits, general-purpose timers, ADC converter, and the very important topic of interrupt structure. Steps are given to show how the timer and external interrupts can be configured.

2.3 Exercises

1. Describe the memory structure of the PIC32 series of microcontrollers. What is the difference between physical and virtual addresses?
2. Explain the functions of the boot memory in a PIC32 series of microcontroller.
3. Explain the differences between a general-purpose register (GPR) and a special function register (SFR).
4. Explain the various ways that the PIC32 series of microcontrollers can be reset. Draw a circuit diagram to show how an external push-button switch can be used to reset the microcontroller.
5. Describe the various clock sources that can be used to provide clock to a PIC32 series of microcontroller. Draw a circuit diagram to show how a 10 MHz crystal can be connected to the microcontroller.
6. Explain how an external clock can be used to provide clock pulses to a PIC32 series of microcontroller.
7. What are the registers of a typical I/O port in a PIC32 series of microcontroller? Explain the operation of the port by drawing the port block diagram.
8. Explain the differences between the PORT and LAT registers.
9. Explain the structure of the ADC converter of a PIC32 series of microcontroller.
10. An LM35DZ-type analogue temperature sensor is connected to analogue port AN0 of a PIC32MX460F512L microcontroller. The sensor provides an analogue output voltage proportional to the temperature, that is, $V_o = 10$ mV/°C. Show the steps required to read the temperature.
11. Explain the differences between a priority interrupt and a nonpriority interrupt.
12. Explain the differences between a single vector interrupt and a multivector interrupt.
13. Show the steps required to configure Timer 1 to generate an interrupt. Assuming a 10 MHz clock frequency, what will be loaded into register PR1 to generate an interrupt every second?
14. Show the steps required to set up external interrupt input INT0 to generate interrupts on its falling edge.
15. Show the steps required to set up Timer 1 to generate interrupts every millisecond having a priority of 7, and subpriority 3.
16. In an application, the CPU registers have been configured to accept interrupts from external sources INT0, INT1, and INT2. An interrupt has been detected. Explain how you can find the source of the interrupt.

CHAPTER 3

PIC32 Microcontroller Development Tools

Chapter Outline

The development of a microcontroller-based system is a complex process. Development tools are hardware and software tools that help programmers to develop and test systems in a relatively short time.

Developing software and hardware for microcontroller-based systems involves the use of editors, assemblers, compilers, debuggers, simulators, emulators, and device programmers.

PIC32 Microcontrollers and the Digilent chipKIT. 978-0-08-099934-0
http://dx.doi.org/10.1016/B978-0-08-099934-0.00003-X

61

A typical development cycle starts with writing the application program using a text editor. The program is then translated into the executable code by using an assembler or a compiler. If the program consists of several modules, these are combined together into a single application program using a linker. At this stage, any syntax errors are detected by the assembler or the compiler and they have to be corrected before an executable code can be generated. In the next stage of the development cycle, a simulator can be used to test the application program without the actual hardware. Simulators can be useful to test the correctness of an algorithm or a program with limited or no input–outputs. Most of the errors can be removed during the simulation. When the program seems to be working, the next stage of the development cycle is to load the executable code to the target microcontroller chip using a device programmer and then test the overall system logic. During this cycle, software and hardware tools such as in-circuit debuggers or in-circuit emulators can be used to analyse the operation of the program, and to display the variables and registers in real time by setting breakpoints in the program.

Development tools for microcontrollers can be classified into two categories: software and hardware. There are many such tools, and the discussion of all these tools is beyond the scope of this book. In this chapter, the commonly used tools are reviewed briefly.

3.1 Software Development Tools

Software development tools are basically computer programs, and they usually run on personal computers, helping the programmer (or system developer) to create and/or modify or test applications programs. Some of the commonly used software development tools are the following:

- Text editors
- Assemblers/compilers
- Simulators

3.1.1 Text Editors

A text editor is a program that allows us to create or edit programs and text files. Windows operating system is distributed with a text editor program called *Notepad*. Using the Notepad, we can create a new program file, modify an existing file, or display or print the contents of a file. It is important to realise that programs used for word processing, such as the *Word*, cannot be used as a text editor. This is because word processing programs are not true text editors as they embed word formatting characters such as bold, italic, underline, etc., inside the text.

Most assemblers and compilers have built-in text editors. Using these editors, we can create our program and then assemble or compile it without having to exit from the editor environment. These editors also provide additional features, such as automatic keyword highlighting, syntax checking, parenthesis matching, comment line identification, and so on. Different parts of a program can be shown in different colours to make the program more readable. For example, comments can be shown in one colour, keywords in another colour, conditional statements in a different colour, and so on. These features can speed up the program development process since most syntax errors can be eliminated during the programming stage.

3.1.2 Assemblers and Compilers

Assemblers generate executable code from assembly language programs. The generated code is usually loaded into the flash program memory of the target microcontroller.

Similarly, compilers generate executable code from high-level language programs. Some of the commonly used compilers for the PIC32 microcontrollers are BASIC, C, PASCAL, MPIDE, and C++.

Assembly language is used in applications where the processing speed is very critical and the microcontroller is required to respond to external and internal events in the shortest possible time. The main disadvantage of assembly language is that it is difficult to develop complex programs using this language. Also, assembly language programs cannot be maintained easily. High-level languages, on the contrary, are easier to learn and complex programs can be developed and tested in a much shorter time. The maintenance of high-level programs is also much easier than the maintenance of assembly language programs.

3.1.3 Simulators

A simulator is a computer program that runs on a PC without any microcontroller hardware, and it simulates the behaviour of the target microcontroller by interpreting the user program instructions using the target microcontroller instruction set. Simulators can display the contents of registers, memory, and the status of input–output ports of the target microcontroller as the user program is interpreted. The user can set breakpoints to stop execution of the program at desired locations and then examine the contents of various registers at the breakpoint. In addition, the user program can be executed in a single-step mode and the memory and registers can be examined as the program executes a single instruction each time a key is pressed. One problem associated with standard simulators is that they are only software tools and any hardware interface is not simulated.

Most microcontroller language development tools also incorporate some form of simulators.

3.1.4 High-Level Language Simulators

These are also known as source-level debuggers, and, like simulators, they are programs that run on a PC. A source-level debugger allows us to find the errors in our high-level programs.

We can set breakpoints in high-level statements, execute the program up to the breakpoint, and then display the values of program variables, the contents of registers, and memory locations at the breakpoint. For example, we can stop a program execution and examine (or modify) the contents of an array.

A source-level debugger can also invoke hardware-based debugging activity using a hardware debugger device. For example, the user program on the target microcontroller can be stopped and the values of various variables and registers can be examined.

3.1.5 Simulators With Hardware Simulation

Some simulators (e.g., Labcenter Electronics VSM, http://www.labcenter.com) incorporate hardware simulation options where various software simulated hardware devices can be connected to microcontroller I/O pins. For example, an electromagnetic motor software module can be connected to an I/O port and its operation can be simulated in software. Although the hardware simulation does not simulate the actual device exactly, it is very useful during early project development cycle.

3.1.6 Integrated Development Environment (IDE)

IDEs are powerful PC-based programs that include everything to edit, assemble, compile, link, simulate, source-level debug, and download the generated executable code to the physical microcontroller chip (using a programmer device). These programs are in the form of graphical user interface (GUI) where the user can select various options from the program without having to exit the program. IDEs can be extremely useful during the development phases of microcontroller-based systems. Most PIC32 high-level language compilers are in the form of an IDE, thus enabling the programmer to do most tasks within a single software development tool.

3.2 Hardware Development Tools

There are numerous hardware development tools available for the PIC32 microcontrollers. Some of these products are manufactured by Microchip Inc., and some by third-party companies. The popular hardware development tools are the following:

- Development boards
- Device programmers
- In-circuit debuggers
- In-circuit emulators
- Breadboards

3.2.1 Development Boards

The development boards are invaluable microcontroller development tools. Simple development boards contain just a microcontroller and the necessary clock circuitry. Some sophisticated development boards contain LEDs, LCDs, push-buttons, serial ports, USB ports, power supply circuits, device programming hardware, etc.

In this section, we shall be looking at the specifications of some of the commercially available PIC32 microcontroller development boards.

PIC32 Starter Kit

This board (see Figure 3.1) is manufactured by Microchip Inc. and can be used in PIC32 microcontroller-based project development. The kit includes everything to write, compile, program, debug, and execute a program.

The kit contains the PIC32 Starter Kit board and a USB Mini-B cable.

The board contains the following:

- PIC32MX360F512L 32-bit microcontroller
- Regulated power supply (+3.3 V) for powering via the USB port
- Processor running at 72 MHz
- On-board debugging
- Debug and power on LEDs
- Three push-button switches for user inputs
- Three LED indicators
- Connectors for I/O ports
- Interface to the I/O expansion board

Figure 3.1: PIC32 Starter Kit

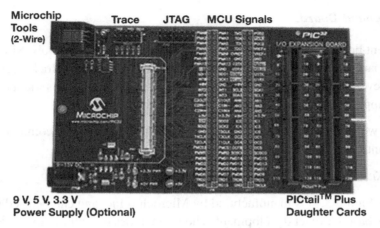

Figure 3.2: I/O Expansion Board

A total of 35 example programs are provided with the kit. Users can download a free MPLAB C32 compiler with limited functionality from company's web site and use the compiler in their projects.

The I/O expansion board (see Figure 3.2) provides full access to all the microcontroller I/O signals. Additional daughter boards can be attached to the expansion board for added functionality.

Microstick II

Microstick II from Microchip Inc. is a small (the size of a stick of gum), low-cost development board (see Figure 3.3), designed for small applications with a few I/O port requirements. The kit is supplied with the following:

- Microstick II board
- USB cable
- PIC32MX250F128 microcontroller (in addition, PIC24 and dsPIC33 microcontrollers are also included)
- Integrated USB programmer/debugger

Figure 3.3: Microstick II

Figure 3.4: PIC32 USB Starter Kit II

- LED and reset button
- Pin headers for I/O access

The board is distributed with free demo programs.

PIC32 USB Starter Kit II

The PIC32 USB Starter Kit II (see Figure 3.4) is a low-cost PIC32 microcontroller development board with USB and CAN functionality. Users can develop USB- and CAN-based applications easily using this kit.

The kit has the following features:

- PIC32MX795F512L 32-bit microcontroller
- On-board crystal
- USB for on-board programming/debugging
- Three push-button switches for user inputs
- Three LED indicators
- Debug and power LEDs
- Regulated power supply
- I/O connector for various expansion boards

PIC32 Ethernet Starter Kit

The PIC32 Ethernet Starter Kit (see Figure 3.5) is a low-cost 10/100 Ethernet development kit manufactured by Microchip Inc., using the PIC32 microcontroller.

Figure 3.5: PIC32 Ethernet Starter Kit

The board has the following features:

- PIC32MX795F512L 32-bit microcontroller
- 32-Bit microcontroller for on-board programming/debugging
- On-board crystal
- Ethernet oscillator
- Three push-button switches for user input
- Three LED indicators
- Debug and power supply LEDs
- RJ-45 Ethernet port
- Connector for various expansion boards

Cerebot MX3cK

The Cerebot MX3cK (see Figure 3.6) is a 32-bit microcontroller development board based on PIC32MX320F128H, and is manufactured by Digilent (www.digilentinc.com). The kit is of low cost and contains everything needed to start developing embedded applications based on 32-bit PIC microcontrollers using the MPIDE IDE. In order to use MPLAB IDE, a programming/debugging device is required.

The kit has the following features:

- PIC32MX320F128H 32-bit microcontroller
- An 80 MHz maximum operating frequency
- Forty-two I/O pins
- Twelve analogue inputs
- Programmed using MPIDE or MPLAB IDE

Figure 3.6: Cerebot MX3cK

- Pmod headers for I/O signals
- I²C connector
- Powered via a USB port or using an external supply

Figure 3.7 shows the functional blocks of the Cerebot MX3cK development board.

Cerebot MX4cK

The Cerebot MX4cK (see Figure 3.8) is a more advanced version of the Cerebot MX3cK with a bigger printed circuit board (PCB) area and more functionality.

The kit has the following features:

- PIC32MX460F512L 32-bit microcontroller
- Pmod headers for I/O ports
- 2 × I²C ports
- 1 × SPI port
- 8 × servo ports
- USB debugging/programming port (for MPLAB IDE)
- USB port for debugging/programming (for MPIDE IDE)

Figure 3.9 shows the functional blocks of the Cerebot MX4cK development board.

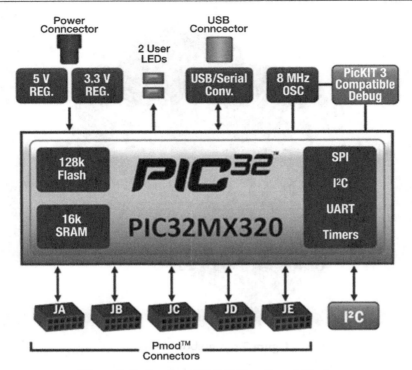

Figure 3.7: Cerebot MX3cK Functional Blocks

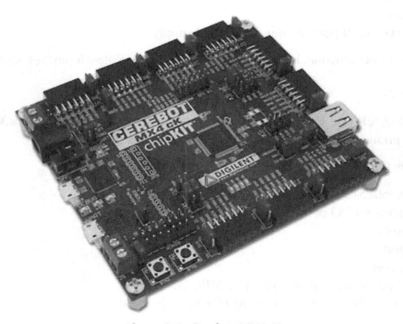

Figure 3.8: Cerebot MX4cK

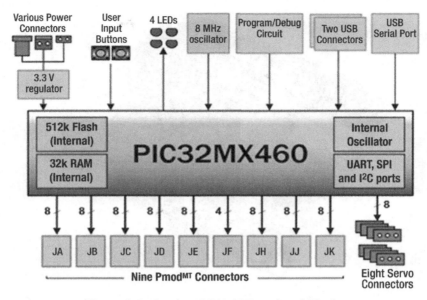

Figure 3.9: Cerebot MX4cK Functional Blocks

Cerebot MX7cK

The Cerebot MX7cK (see Figure 3.10) is the most advanced version of the Digilent MX series of 32-bit development boards.

The kit has the following features:

- PIC32MX795F512L 32-bit microcontroller
- RJ-45 Ethernet port
- 2 × I²C ports
- 1 × SPI port
- 2 × CAN ports
- 2 × SPI/UART ports
- 1 × USB UART port
- 2 × USB ports
- Pmod headers for I/O pins

Figure 3.11 shows the functional blocks of the Cerebot MX7cK development board.

MINI-32 Board

This is a small development board (see Figure 3.12) manufactured by mikroElektronika (www.mikroe.com) that contains a PIC32MX534F064H 32-bit microcontroller. The board operates with 3.3 V power supply, and on-board regulator allows the board to be powered from a USB port.

Figure 3.10: Cerebot MX7cK

The features of this board are the following:

- PIC32MX534F064H 32-bit microcontroller
- On-board crystal
- I/O pins at the edges

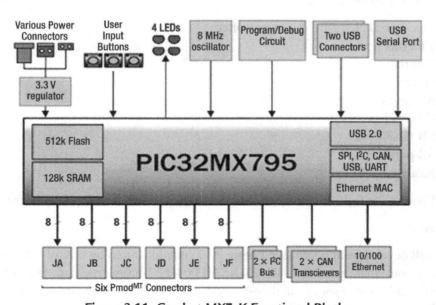

Figure 3.11: Cerebot MX7cK Functional Blocks

Figure 3.12: MINI-32 Board

- Supports CAN communication
- Comes with preprogrammed USB BootLoader
- Fully supported by mikroElektronika mikroC PRO for PIC32 compiler

EasyPIC Fusion V7

The EasyPIC Fusion V7 (see Figure 3.13) combines support for three 16- and 32-bit different
PIC microcontroller architectures – dsPIC33, PIC24, and PIC32 – in a single development board.

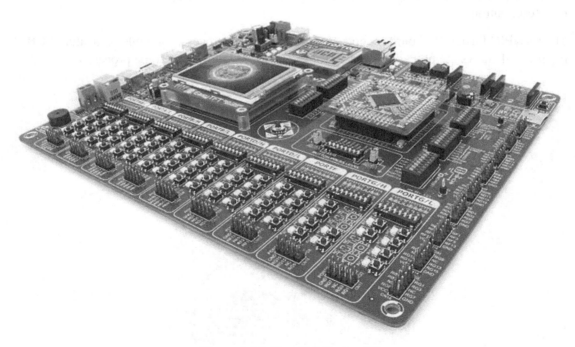

Figure 3.13: EasyPIC Fusion V7

The board has the following features:

- PIC32MX795F512L 32-bit microcontroller module
- On-board programmer (mikroProg)
- On-board in-circuit debugger (mikroICD)
- Sixty-eight push-button switches
- Sixty-eight LED indicators
- CAN support
- USB support
- Piezo buzzer
- LM35/DS1820 temperature sensor sockets
- RJ-45 Ethernet connector
- I²C EEPROM
- Serial flash memory
- Stereo MP3 codec
- 2 × mikroBUS sockets
- Audio in and out jack sockets
- mikroSD card slot
- TFT colour display socket
- I/O port headers
- 3.3 V power supply (can be powered from USB or external supply)
- Reset button

The EasyPIC Fusion V7 board accepts an external plug-in processor module on a small PCB. Figure 3.14 shows the processor module for the PIC32MX460F512L-type processor.

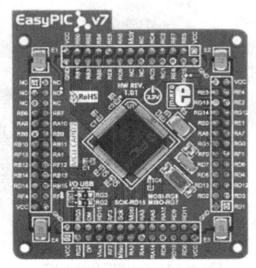

Figure 3.14: PIC32MX460F512L Processor Module

Figure 3.15: Mikromedia for PIC32 Board

Mikromedia for PIC32

The mikromedia for PIC32 board (see Figure 3.15) is a small board with an integrated touchscreen TFT colour display. In addition, the board contains a stereo MP3 codec chip and a microSD card slot. The device can be powered from an external USB port or from an external battery. A preprogrammed BootLoader program enables the microcontroller chip to be programmed. The device is reset using a reset button.

Multimedia for PIC32MX7

This development board includes a PIC32MX795F512L 32-bit microcontroller and is used for multimedia-based applications. A large touchscreen colour TFT display is provided (see Figure 3.16) with on-board push-button switches for game applications. In addition, an Ethernet interface and a microSD card slot are available to store images or data.

Olimex PIC32 Development Board

This is a low-cost 32-bit microcontroller development board (see Figure 3.17) with a high-performance PIC32MX460F512L microcontroller (http://www.olimex.com).

Figure 3.16: Multimedia for PIC32MX7 Board

Figure 3.17: Olimex PIC32 Development Board

The board offers the following features:

- PIC32MX460F512L microcontroller
- Audio input and output
- USB interface
- SD card slot
- JTAG connector
- 84 × 84 pixel LCD
- On-board crystal
- Joystick
- Reset button
- 3.3 V voltage regulator
- I/O pins on connectors
- Development PCB area

PIC32-MAXI-WEB Development Board

This board (see Figure 3.18) from Olimex features a PIC32 microcontroller with an embedded 100 Mbit Ethernet module. A large 240 × 320 TFT touchscreen LCD is provided with the board for graphical applications.

Figure 3.18: PIC32-MAXI-WEB Board

This board has the following features:

- PIC32MX795F512L 32-bit microcontroller
- 320 × 240 LCD
- 2 × opto-isolated digital inputs
- 2 × CAN interface
- Accelerometer sensor
- Temperature sensor
- microSD card slot
- 2 × relays
- RS232 interface
- 3 × LED indicators
- Reset button
- 3.3 V voltage regulator

LV-32MX V6

The LV-32MX V6 (see Figure 3.19) is a PIC32 development system manufactured by mikroElektronika (www.mikroe.com), and is equipped with many on-board modules, including multimedia peripherals that give great power and flexibility for system development. This development board is fully compatible with the mikroC PRO for PIC32 compiler.

Figure 3.19: LV-32MX V6

The board offers the following features:

- PIC32MX460F512L
- Eighty-five push-button switches
- Eighty-five LEDs
- SD card slot
- Reset button
- Power supply regulator
- Colour TFT display with touchscreen
- CAN support
- On-board programmer and debugger
- Serial EEPROM
- Serial flash memory
- Stereo codec chip
- Chip-on-Glass (COG) LCD
- 2 × UART connectors
- DS1820 temperature sensor socket
- I/O headers

3.2.2 Device Programmers

After writing and translating the program into executable code, the resulting HEX file should be loaded to the target microcontroller program memory. Device programmers are used to load the program memory of the actual microcontroller chip. The type of device programmer to be used depends on the type of microcontroller to be programmed. For example, some device programmers can program only PIC16 series, some can program both PIC16 and PIC18 series, and some are used to program different models of microcontrollers (e.g., Intel 8051 series).

As we have seen in the previous section, some microcontroller development kits include on-board device programmers, and thus there is no need to remove the microcontroller chip and insert into the programming device. In this section, some of the popular device programmers that can be used to program PIC32 series of microcontrollers are described.

mikroProg

mikroProg (see Figure 3.20) is a small handheld programmer manufactured by mikroElektronika and supports all PIC microcontrollers from PIC10 and PIC12 to PIC16, PIC18, dsPIC, PIC24, and PIC32.

mikroProg programmer is supported by all the compilers of the company. The device is connected to a PC via a USB cable, and to the target development system. The microcontroller in the target system is programmed by first compiling and sending the HEX code to the programmer device.

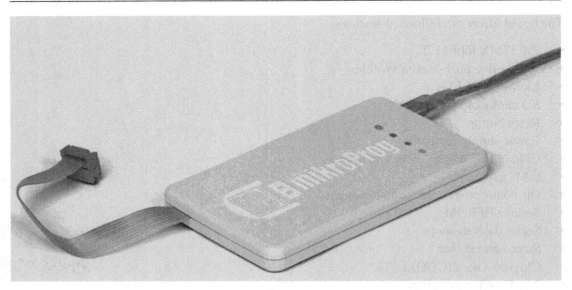

Figure 3.20: mikroProg Device Programmer

3.2.3 In-Circuit Debuggers

An in-circuit debugger is a hardware device, connected between a PC and the target microcontroller test system, and is used to debug real-time applications faster and easier. With in-circuit debugging, a monitor program runs in the PIC microcontroller in the test circuit. The programmer can set breakpoints on the PIC, run code, single step the program, examine variables and registers on the real device, and, if required, change their values. An in-circuit debugger uses some memory and I/O pins of the target PIC microcontroller during the debugging operations. With some in-circuit debuggers, only the assembly language programs can be debugged. Some more powerful debuggers enable high-level language programs to be debugged.

In-circuit debuggers also include programming functions that enable the target microcontroller to be programmed. Some of the popular in-circuit debuggers are PICkit 3, ICD3, and Real Ice from Microchip (www.microchip.com), and mikroProg from mikroElektronika. These devices can be used with all types of PIC microcontrollers.

3.2.4 In-Circuit Emulators

The in-circuit emulator is one of the oldest and the most powerful methods of debugging a microcontroller system. In fact, it is the only tool that substitutes its own internal processor for the one in your target system. Like all in-circuit debuggers, the emulator's most fundamental resource is target access – the ability to examine and change the contents of registers, memory, and I/O. However, since the ICE replaces the CPU, it generally does not require working CPU on the target system to provide this capability. This makes the in-circuit emulator by far the best tool for troubleshooting new or defective systems. Usually every microcontroller family has its own set of in-circuit emulator. For example, an in-circuit emulator for the PIC16 microcontrollers cannot be used for the PIC18 microcontrollers. Because of this, in order to lower the costs, emulator manufacturers provide a multiboard solution to in-circuit emulation. Usually a base board is provided that is common to most microcontrollers in the family. For example, the same base board can be used by all PIC microcontrollers. Then, probe cards are available for individual microcontrollers. When it is required to emulate a new microcontroller in the same family, it is sufficient to purchase just the probe card for the required microcontroller.

3.2.5 Breadboard

When we are building an electronic circuit, we have to connect the components as shown in Figure 3.21. This task can usually be carried out on a stripboard or a PCB by soldering the components together. The PCB approach is used for circuits that have been tested and that function as desired and also when the circuit is to be made permanent. It is not economical to make a PCB design for one or only a few applications.

During the development stage of an electronic circuit, it may not be known in advance whether or not the circuit will function correctly when assembled. A solderless breadboard is then usually used to assemble the circuit components together. A typical breadboard is shown in Figure 3.21. The board consists of rows and columns of holes

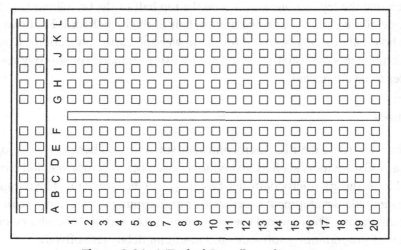

Figure 3.21: A Typical Breadboard Layout

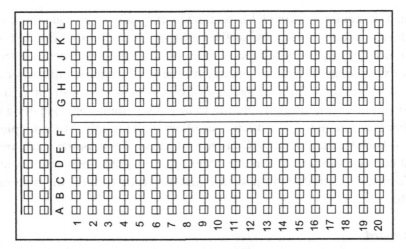

Figure 3.22: Internal Wiring of the Breadboard Shown in Figure 3.21

that are spaced so that integrated circuits and other components can be fitted inside them. The holes have spring actions so that the component leads can be held tight inside the holes. There are various types and sizes of breadboards depending on the complexity of the circuit to be built. The boards can be stacked together to make larger boards for very complex circuits. Figure 3.22 shows the internal connection layout of the breadboard given in Figure 3.21.

The top and bottom half parts of the breadboard are separate with no connection between them. Columns 1–20 in rows A–F are connected to each other on a column basis. Similarly, rows G–L in columns 1–20 are connected to each other on a column basis. Integrated circuits are placed such that the legs on one side are on the top half of the breadboard, and the legs on the other side of the circuit are on the bottom half of the breadboard. First two columns on the left of the board are usually reserved for the power and earth connections. Connections between the components are usually carried out by using stranded (or solid) wires plugged inside the holes to be connected.

3.3 Summary

This chapter has described the PIC microcontroller software and hardware development tools. It is shown that software tools such as text editors, assemblers, compilers, and simulators may be useful tools during microcontroller-based system development. The required useful hardware tools include development boards/kits, programming devices, in-circuit debuggers, and in-circuit emulators. The required useful software tools include assemblers, compilers, simulators, device programming software, and in-circuit debugger software.

3.4 Exercises

1. Describe the various phases of the microcontroller-based system development cycle.
2. Give a brief description of the microcontroller development tools.
3. Explain the advantages and disadvantages of assemblers and compilers.
4. Explain why a simulator can be a useful tool during the development of a microcontroller-based product.
5. Explain in detail what a device programmer is. Give an example device programmer for the PIC32 series of microcontrollers.
6. Describe briefly the differences between in-circuit debuggers and in-circuit emulators. List the advantages and disadvantages of each type of debugging tool.

CHAPTER 4

The Cerebot MX3cK (chipKIT MX3) Development Board

Chapter Outline

This book is about using the Cerebot series of 32-bit microcontroller development boards in projects. The low-cost and highly popular MX3cK development board (name has been changed to chipKIT MX3) will be used in the projects in this book. In this chapter, the basic hardware details of this popular development board will be given in detail.

85

4.1 The chipKIT MX3 Development Board

The chipKIT MX3 development board is based on the Microchip PIC32MX320F128H microcontroller, which is a member of the PIC32 microcontroller family. This development board is compatible with Digilent's Pmod hardware peripheral modules, and is suitable for use with the Microchip MPLAB IDE software tools and chipKIT MPIDE development environment. Thus, a large number of existing Arduino-compatible examples, libraries, and reference resources are available for use with this development board.

The chipKIT MX3 provides 42 I/O pins out of which 11 can be used as analogue inputs in addition to their use as general-purpose digital I/O pins. A number of peripheral functions, such as SPI, I²C, UART, and PWM outputs, are supported. The development board can be powered via its USB port. In addition, external power supply (AC/DC power adapter) or batteries can be used to power the board.

Further details about the chipKIT MX3 board can be obtained from the *chipKIT MX3 Board Reference Manual*.

The basic specifications of the chipKIT MX3 development board are as follows:

• PIC32MX320F128H microcontroller
• A total of 128k flash memory
• A total of 16k RAM memory
• A total of 80 MHz maximum operating frequency
• Pmod peripheral connectors
• 3.3 V operation
• Forty-two I/O pins
• Eleven analogue input pins (0–3.3 V)
• A total of 75 mA operating current

Figure 4.1 shows the chipKIT MX3 development board with the functions of various components labelled on the board.

A functional block diagram of the chipKIT MX3 development board is shown in Figure 4.2.

4.1.1 Power Supply

Figure 4.3 shows a schematic of the power supply arrangement. The chipKIT MX3 can be powered either from USB port or from an external power supply. If both supplies are used, then the external supply is selected automatically. Two voltages are used on the board: +5 V (VCC5V0 bus) and +3.3 V (VCC3V3 bus).

External +5 V supply can be directly applied to the system by placing jumper JP2 in position BYP. Alternatively, 7–20 V external DC supply can be applied to the board with jumper

Figure 4.1: The chipKIT MX3 Development Board

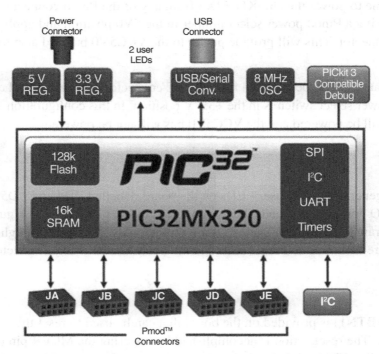

Figure 4.2: Functional Block Diagram of the chipKIT MX3 Development Board

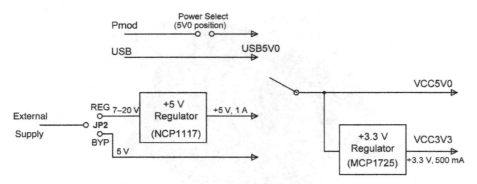

Figure 4.3: Schematic of the Power Supply Arrangement

JP2 in the REG position. In this configuration, the voltage regulator NCP1117 generates the required +5 V supply for the board, with the maximum current capacity of 1 A. The on-board +5 V supply is labelled as VCC5V0.

The +3.3 V supply is generated by the voltage regulator MCP1725, with the maximum current capacity of 500 mA. The input to this regulator must not exceed +6 V. The on-board +3.3 V supply is labelled as VCC3V3.

It is also possible to power the chipKIT MX3 from any of the Pmod connectors. This can be done by placing a Pmod power select jumper in the 5V0 position and applying +5 V through the connector. This will provide power to the VCC5V0 bus and also to the +3.3 V regulator.

The VCC3V3 bus can also be powered from the I²C connector or from a Pmod connector when the power select switch is in the +3.3 V position. In this configuration, only the VCC3V3 bus will be powered and the VCC5V0 bus will not be powered.

4.1.2 LEDs

There are two general-purpose user LEDs on the board, labelled LD4 and LD5. LD4 is connected to I/O port pin RF0, and LD5 is connected to I/O port pin RF1. Figure 4.4 shows the circuit diagram of the LEDs. The LEDs are turned on by driving them high. I/O pins RF0 and RF1 are dedicated to these LEDs and are not available at any connector.

4.1.3 Reset

A reset button (BTN1) is provided on the board that can be used to reset the microcontroller. The reset action is accomplished by lowering the MCLR pin of the microcontroller. Figure 4.5 shows the circuit diagram of the reset circuit.

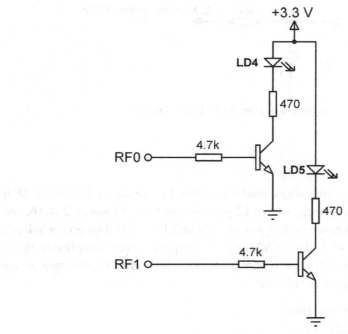

Figure 4.4: Circuit Diagram of the LEDs

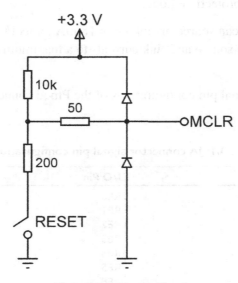

Figure 4.5: Reset Circuit

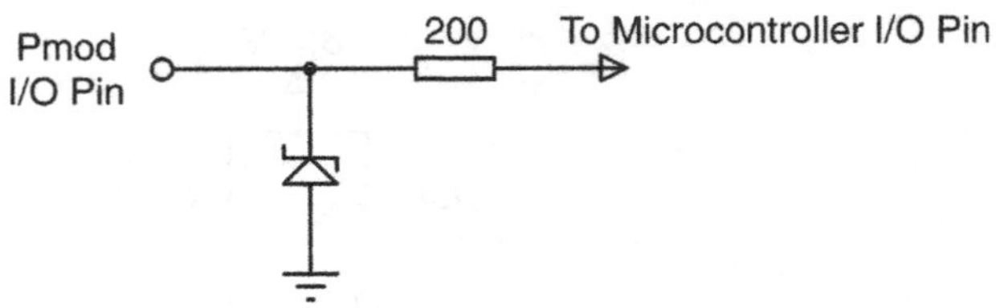

Figure 4.6: Pmod Pin Connections

4.1.4 Pmod Connectors

The chipKIT MX3 has five connectors called Pmod (see Figure 4.1), labelled JA–JF for peripheral modules. These connectors have 12 pins organised in 2 rows as 2 × 16, and each connector has an associated power select jumper labelled JPA–JPF. The power select jumpers are used to select (between +5 and +3.3 V) the power supply voltage supplied to the power supply pin on the Pmod connector. Each connector provides eight I/O pins, two power pins, and two ground pins, organised as follows:

- Pins 1–4 and 7–10 are signal pins.
- Pins 5 and 11 are ground pins.
- Pins 6 and 12 are power pins.

The upper six pins are numbered 1–6 left to right as viewed from the top of the board. The lower six pins are numbered 7–12 left to right.

As shown in Figure 4.6, each Pmod pin is connected to a microcontroller I/O pin through a 200 Ω series resistor and a protection diode.

The PIC32 microcontroller can source or sink up to 18 mA on its I/O pins. However, it is recommended to keep the source and sink currents to a maximum of 7 and 12 mA, respectively.

Tables 4.1–4.5 show the signal pin configurations of the Pmod connectors.

Table 4.1: JA connector signal pin configuration.

Pmod Connector	I/O Pin
JA1	RE0
JA2	RE1
JA3	RE2
JA4	RE3
JA7	RE4
JA8	RE5
JA9	RE6
JA10	RE7

Table 4.2: JB connector signal pin configuration.

Pmod Connector	I/O Pin
JB1	RD9
JB2	RF3
JB3	RF2
JB4	RF6
JB7	RD6
JB8	RD5
JB9	RD4
JB10	RD7

Table 4.3: JC connector signal pin configuration.

Pmod Connector	I/O Pin
JC1	RB8
JC2	RF5
JC3	RF4
JC4	RB14
JC7	RB0
JC8	RB1
JC9	RD0
JC10	RD1

Table 4.4: JD connector signal pin configuration.

Pmod Connector	I/O Pin
JD1	RB2
JD2	RD2
JD3	RD10
JD4	RB9
JD7	RB12
JD8	RD3
JD9	RD11
JD10	RB13

Table 4.5: JE connector signal pin configuration.

Pmod Connector	I/O Pin
JE1	RG9
JE2	RG8
JE3	RG7
JE4	RG6
JE7	RD8
JE8	RB5
JE9	RB4
JE10	RB3

4.1.5 CPU Clock

As shown in Figure 4.7, the microcontroller on the board is operated from an 8 MHz crystal. However, the internal phase-locked-loop (PLL) module of the microcontroller can be programmed to provide operating system clock (SYSCLK) frequency up to 80 MHz. The peripheral clock (PBCLK) frequency can be programmed as a division of the system clock frequency (factors of 1, 2, 4, or 8), and its maximum value can be 10 MHz.

4.1.6 I^2C Bus Interface

The I^2C bus on the board is an open-collector bus with medium speed communication, providing master and slave operation using either 7- or 10-bit device addressing. Although the PIC32MX320 microcontroller provides two I^2C interfaces, the chipKIT

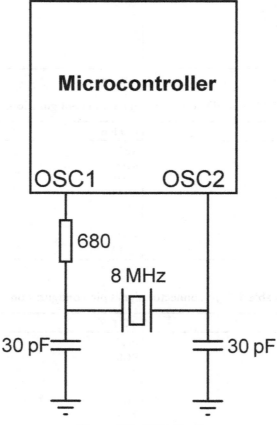

Figure 4.7: CPU Clock

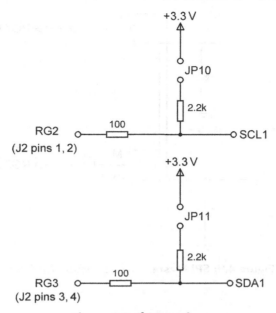

Figure 4.8: I²C Interface

MX3 is designed to provide only one interface (I²C1; the second I²C interface, I²C2, can be accessed at pins 2 and 3 of Pmod connector JC) on connector J2.

As shown in Figure 4.8, 2.2k resistors are used to pull up the SCL1 and SDA1 lines to the power supply (there are no pull-up resistors on the second I²C interface, I²C2).

4.1.7 SPI Bus Interface

The PIC32 microcontroller supports two SPI bus interfaces. Each interface has four signals, named SS (Slave Select), MOSI (Master Out Slave In), MISO (Master In Slave Out), and SCK (Serial Clock). SPI1 supports master mode only and is accessed via Pmod connector JB. SPI2 is accessed via Pmod connector JE:

JB-01: SS1
JB-02: MOSI1
JB-03: MISO1
JB-04: SCK1

JE-01: SS2
JE-02: MOSI2
JE-03: MISO2
JE-04: SCK2

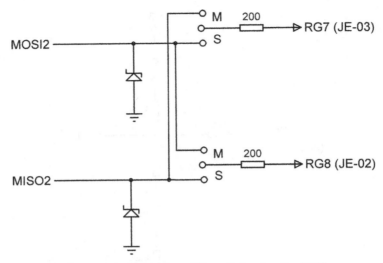

Figure 4.9: SPI Master/Slave Selection for SPI2

As shown in Figure 4.9, jumpers JP6 and JP8 are used to select between master and slave for the SPI2 bus interface.

4.1.8 UART Interface

The PIC32MX320 microcontroller provides two UART interfaces, named UART1 and UART2. Either two-wire (TX and RX) or four-wire (RX, TX, RTS, CTS) communication is possible. UART1 and UART2 can be accessed from Pmod connectors JB and JC, respectively:

    JB-01: UART1 (CTS)
    JB-02: UART1 (TX)
    JB-03: UART1 (RX)
    JB-04: UART1 (RTS)

    JC-01: UART2 (CTS)
    JC-02: UART2 (TX)
    JC-03: UART2 (RX)
    JC-04: UART2 (RTS)

Figure 4.10 shows the UART1 schematic. Notice that the handshaking signals CTS, DSR, DCD, and RI are available on connector J3.

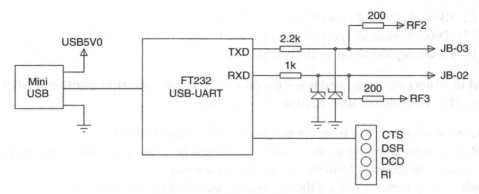

Figure 4.10: UART1 Schematic

4.1.9 Analogue Inputs

The chipKIT MX3 development board provides 11 analogue inputs (A0–A10). Interface to these analogue inputs is through Pmod connectors JC, JD, and JE:

A0: JC-01 (RB8)
A1: JC-04 (RB14)
A2: JC-07 (RB0)
A3: JC-08 (RB1)
A4: JD-01 (RB2)
A5: JD-04 (RB9)
A6: JD-07 (RB12)
A7: JD-10 (RB13)
A8: JE-08 (RB5)
A9: JE-09 (RB4)
A10: JE-10 (RB3)

The internal reference voltages for the analogue inputs can be selected as +3.3 V.

4.1.10 External Interrupts

PIC32 microcontroller supports five external interrupts named INT0–INT4. When using the MPIDE integrated development environment, these external interrupts are available in the following Pmod connectors:

INT0: JB-04, logical I/O port 11 (port pin RF6)
INT1: JE-07, logical I/O port 36 (port pin RD8)

INT2: JB-01, logical I/O port 8 (port pin RD9)
INT3: JD-03, logical I/O port 26 (port pin RD10)
INT4: JD-09, logical I/O port 39 (port pin RD11)

External interrupts are configured using the **attachInterrupt(interrupt number, ISR, mode)** function. The arguments of this function are as follows:

Interrupt number: The interrupt number is 0 (INT0) to 4 (INT4).

ISR: This is the interrupt service routine. This must be a function with no arguments, and no data can be returned from the interrupt service routine.

Mode: This parameter defines the interrupt mode. Valid modes are as follows:

LOW	Interrupt triggered when the pin is LOW
CHANGE	Interrupt triggered when the pin changes value
RISING	Interrupt triggered when the pin state changes from LOW to HIGH
FALLING	Interrupt is triggered when the pin state changes from HIGH to LOW
HIGH	Interrupt is triggered when the pin is HIGH

Function **detachInterrupt(interrupt number)** is used to turn off an interrupt.

4.1.11 Board Connectors and Jumpers

A list of the various connectors and jumpers on the chipKIT MX3 development board and the functions of these jumpers are given in Table 4.6.

Table 4.6: chipKIT MX3 board jumpers.

Jumpers	Description
JA–JE	Pmod connectors
JPA–JPE	Pmod connector power select
J1	Mini USB connector
J2	I²C interface
J3	Additional UART signals
J4	External power connector
JP1	Used to disconnect the USB from the MCLR reset input. Must be installed to use the board with the MPIDE development software
JP2	External power select. If the jumper is in RTEG position, then external voltage is routed through the 5 V regulator
JP3	Microchip development connector
JP6, JP8	SPI2 master/slave select
JP10, JP11	I²C pull-up resistor enable

4.2 Pmod Peripheral Modules

Digilent offers a large number of Pmod-based peripheral devices in the form of small I/O interface boards for project development using their development boards. Pmod boards include sensors, data acquisition, simple I/O, LEDs, seven-segment displays, GPS, accelerometers, external memory, D/A converters, audio amplifiers, and many more.

Some of the popular Pmod modules are described briefly in this section. Detailed description of the Pmod modules can be found on the Digilent web site (www.digilentinc.com).

4.2.1 PmodSD – SD Card Slot

This is a small Secure Digital media card module (see Figure 4.11) that enables the programmer to use SD cards in projects.

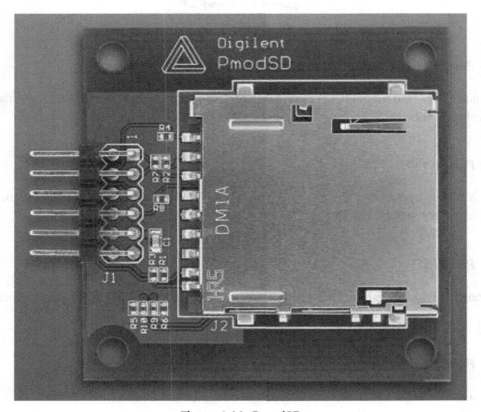

Figure 4.11: PmodSD

Figure 4.12: PmodCLP

4.2.2 PmodCLP – Character LCD With Parallel Interface

This small board (see Figure 4.12) provides 2-line by 16-character (2 × 16) LCD interface to the development board.

4.2.3 PmodKYPD

This is a 16-key keypad (see Figure 4.13) with labelled buttons. Keypads are useful in projects where data entry may be required.

4.2.4 PmodSSD

This board provides two-digit seven-segment display (see Figure 4.14) to the development board.

4.2.5 PmodBTN

This is a small board with four debounced momentary push-button switches (see Figure 4.15).

Figure 4.13: PmodKYPD

Figure 4.14: PmodSSD

Figure 4.15: PmodBTN

4.2.6 PmodSTEP

This is a stepper motor controller board (see Figure 4.16) that can be used with four- and six-pin stepper motors.

4.2.7 PmodTMP3

This is a programmable digital temperature sensor board with an accuracy of ±1°C (see Figure 4.17).

Figure 4.16: PmodSTEP

Figure 4.17: PmodTMP3

4.2.8 PmodDA1

This board incorporates four 8-bit D/A converters (see Figure 4.18).

4.2.9 PmodRTCC

This is a real-time clock/calendar board with lithium coin cell battery backup (see Figure 4.19).

4.2.10 PmodBB

This board contains a wire wrap area and a small breadboard that can be used in projects (see Figure 4.20).

4.2.11 PmodAMP2

This is a small 2.5 W audio amplifier module (see Figure 4.21).

Figure 4.18: PmodDA1

Figure 4.19: PmodRTCC

Figure 4.20: PmodBB

Figure 4.21: PmodAMP2

4.3 Summary

This chapter has explained the hardware details of the chipKIT MX3 development board. The functions and circuit diagrams of various components used on the board have been explained in the chapter. In addition, some of the popular Pmod peripheral modules have been described briefly.

CHAPTER 5

The MPIDE Programming Environment and Programming in C

Chapter Outline

PIC32 Microcontrollers and the Digilent chipKIT. 978-0-08-099934-0
http://dx.doi.org/10.1016/B978-0-08-099934-0.00005-3

105

The Digilent Multi-Platform Integrated Development Environment (MPIDE) is a modified version of the popular Arduino IDE. MPIDE is compatible with both native Arduino hardware and software and with Digilent's chipKIT PIC-32–based family of microcontrollers. The IDE is based on a simplified set of C/C++ language and libraries, with the aim of providing a microcontroller software development environment with less technically minded people.

In this chapter, we shall be looking at the basic features of the MPIDE development environment and see how we can develop simple programs using this language and the environment.

5.1 Installing MPIDE

MPIDE is available free of charge, and it can be downloaded from the Digilent web site. The steps to download and start the MPIDE are given below:

    Step 1: Download MPIDE from the Digilent web site by following the link:
            http://chipkit.net/started/
    Step 2: The distribution files are zipped together, and they should be unzipped in
    a directory, preferably called MPIDE.

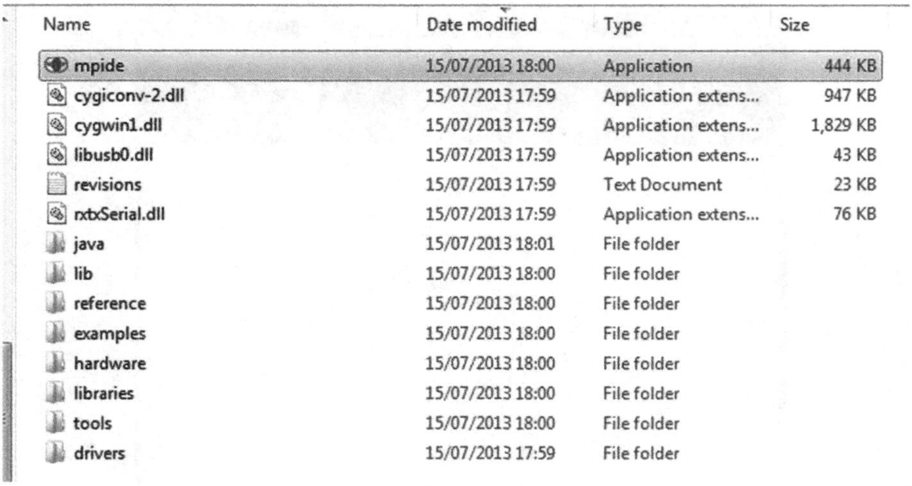

Name	Date modified	Type	Size
mpide	15/07/2013 18:00	Application	444 KB
cygiconv-2.dll	15/07/2013 17:59	Application extens...	947 KB
cygwin1.dll	15/07/2013 17:59	Application extens...	1,829 KB
libusb0.dll	15/07/2013 17:59	Application extens...	43 KB
revisions	15/07/2013 17:59	Text Document	23 KB
rxtxSerial.dll	15/07/2013 17:59	Application extens...	76 KB
java	15/07/2013 18:01	File folder	
lib	15/07/2013 18:00	File folder	
reference	15/07/2013 18:00	File folder	
examples	15/07/2013 18:00	File folder	
hardware	15/07/2013 18:00	File folder	
libraries	15/07/2013 18:00	File folder	
tools	15/07/2013 18:00	File folder	
drivers	15/07/2013 17:59	File folder	

Figure 5.1: Double Click Executable mpide.exe

Step 3: MPIDE can be started by clicking on the executable called mpide.exe in unzipped installation directory (see Figure 5.1). It is advisable to create a shortcut to mpide.exe and place it on the desktop.

When MPIDE is started, the screen shown in Figure 5.2 will be displayed.

5.2 The MPIDE

The MPIDE screen consists of the following parts:

    Menu
    Toolbar with buttons
    Status line
    Output window

The middle part of the screen is used to enter user programs.

5.2.1 The Menu

The menu has the following parts:

    File
    Edit
    Sketch
    Tools
    Help

Figure 5.2: MPIDE Screen

As shown in Figure 5.3, the **File** menu option is used to open an existing file, to create a new file, to look at the examples provided by the IDE, and for other file processing operations such as saving, closing, and printing files.

The **Edit** menu option is used for copy and paste operations and for finding text.

The **Sketch** option is used to verify/compile and to add a file or a library.

The **Tools** menu option is used to select a board, to select a programmer, or to configure the serial port.

The **Help** menu provides help on various aspects of the MPIDE environment.

5.2.2 The Toolbar

The toolbar consists of five buttons with the following functions:

> **Verify**: Compiles and checks the code for errors
> **Upload**: Uploads the compiled code to the target microcontroller chip

Figure 5.3: The File Menu Options

New: Creates a new file
Open: Opens an existing file
Save: Saves an already opened file

5.2.3 The Status Line

The status line just above the bottom of the screen shows the status of the current operation.

5.2.4 The Output Window

The output window displays various messages about the current operation, such as error messages.

5.3 The C LANGUAGE

In this section, we shall be looking at various features of the C programming language including the built-in functions and libraries and see how we can use them in our programs.

Figure 5.4 shows the simplest structure of a C program. This program flashes an LED connected to I/O pin 42 of the chipKIT MX3 development board. Do not worry if you do not understand the operation of the program at this stage as all will be clear as we progress through this chapter. Some of the programming concepts used in Figure 5.4 are described in the following sections.

All programs consist of at least two functions: *setup()* and *loop()*.

The *setup()* function is called automatically when a program starts. This function is used to initialise variables, modes of I/O pins, and various libraries. This function is run only once after each power-up or after the microcontroller is reset.

The *loop()* function is where the main program lies, and the program loops here continuously to perform the required tasks.

```
/*=========================================================================
                            LED FLASHING PROGRAM
                            =====================


This program flashes the LED connected to I/O pin 42 (RF0) of the ChipKIT MX3 development
Board every second.


Author:        Dogan Ibrahim
Date:          May, 2014
File:          LED
Board:         ChipKIT MX3
=========================================================================*/
int ledPin = 42;                          // LED port

void setup()                              // initialization
{
  pinMode(ledPin, OUTPUT);                // set digital pin as output
}


void loop()
{
        digitalWrite(ledPin, HIGH);       // turn LED ON
        delay(1000);                      // wait 1 second
        digitalWrite(ledPin, LOW);        // turn LED OFF
        delay(1000);                      // wait 1 second
}
```

Figure 5.4: Structure of a Simple C Program

5.3.1 Comments

Comments are used by programmers to clarify the operation of the program or a program statement. Comment lines are ignored and not compiled by the compiler and their use is optional. Two types of comments can be used in a C program: long comments and short comments.

Long comments start with characters "/*" and end with characters "*/." These comments are usually used at the beginning of a program to describe what the program does, the type of processor used, brief description of the algorithm, and the interface to any external hardware used. In addition, the name of the author, the date program was written, and the program filename are usually given in long comment lines to help make any future development of the program easier. As shown in Figure 5.4, long comment lines can extend over many lines.

Short comment lines start with characters "//" and are usually used at the end of statements to describe the operation performed by the statements. These comments can be inserted at any other places of programs as well, for example:

```
//
// Increment variable Cnt by one
//
Cnt = Cnt + 1;
```

Short comments can occupy only one line, and there is no need to terminate them.

5.3.2 White Spaces

White spaces in programs are blanks, spaces, tabs, and new-line characters. All white spaces are ignored by the C compiler. Thus, the following statements are all identical:

```
char i, j, k;
```

or

```
char i,      j,k;
```

or

```
char i,
    j,  k;
```

or

```
char i,j,k
;
```

5.3.3 Terminating Program Statements

In C programs, all statements must be terminated with the semicolon ";" character; otherwise, a compilation error occurs:

```
char i, j, k        // error
char i, j, k;       // correct
j = b + 2           // error
j = b + 2;          // correct
```

5.3.4 Case Sensitivity

In C programs, all names are case-sensitive with variables with lowercase names being different from those with uppercase names. Thus, the following variables are all different and they represent different locations in memory:

Total total ToTal TotaL TOTAL totaL toTAL TOTAl ToTAL

5.3.5 Variable Names

In C programs, variable names can begin with an underscore character, or with an alphabetical character. Thus, valid variable names can start with characters a–z, A–Z, and the "_" character. Digits 0–9 can be used after valid characters. Variable names must be unique and can extend to up to 31 characters. Examples of valid variable names are:

Total   Total1   Sum5   _Name   SumOfNumbers   UserName   x123

Examples of invalid variable names are:

9Sum   ?Name   +Count   £Pound   45Total   /Address   -cnt

5.3.6 Data Types

The C language supports a large number of variable types. Table 5.1 gives a list of the supported variable types. Examples are given below to show how these variables can be used in programs:

boolean: This is a 1-bit variable that can take values 0 or 1. In the following example, variable flag is declared as a bit and is assigned 1:

```
boolean flag;
flag = 1;
```

Table 5.1: C language data types.

Type	Size (Bits)	Range
unsigned char	8	0–255
unsigned short int	8	0–255
unsigned int	16	0–65,535
unsigned long int	32	0–4,294,967,295
signed char	8	−128 to 127
signed short int	8	−128 to 127
signed int	16	−32,768 to 32,767
signed long int	32	−2,147,483,648 to 2,147,483,647
float	32	±1.17549435082E−38 to ±6.80564774407E38
double	32	±1.17549435082E−38 to ±6.80564774407E38
long double	32	±1.17549435082E−38 to ±6.80564774407E38

Note that variables can be assigned values during their declarations. Thus, the above statement can also be written as:

boolean flag = 1;

unsigned char or byte: These are 8-bit unsigned character or integer variables ranging from 0 to 255 and occupying 1 byte in memory. In the following example, variable **Sum** is assigned value 125:

unsigned char Sum = 125;

or

byte Cnt = 250;

char: These are 8-bit signed character or integer variables ranging from −128 to +127 and occupying 1 byte in memory. In the following example, variable **Cnt** is assigned value −25:

char Cnt;
Cnt = −25;

int: These are 16-bit signed variables ranging from −32,768 to +32,767 and occupying 2 bytes in memory. In the following example, variable **Total** is assigned value −23,000:

int Total = −23000;

unsigned int or word: These are 16-bit unsigned variables ranging from 0 to 65,535 and occupying 2 bytes in memory. In the following example, variable **Sum** is assigned value 33,000:

unsigned int Sum;
Sum = 33000;

or

word Sum = 5000;

long: These are 32-bit signed variables ranging from −2,147,483,648 to +2,147,483,647 and occupying 4 bytes in memory. In the following example, variable **Total** is assigned value −200:

    long Total = −200;

unsigned long: These are 32-bit unsigned variables ranging from 0 to 4,294,967,295 and occupying 4 bytes in memory. In the following example, variable **x1** is assigned value 1250:

    unsigned long x1 = 1250;

float or double: These are floating point numbers in the range $−1.5 \times 10^{45}$ to $+3.4 \times 10^{38}$ occupying 4 bytes in memory. In the following example, variable **Max** is assigned value 12.52:

    float Max = 12.52;

5.3.7 Number Bases

C language supports the decimal, binary, hexadecimal, and octal number bases. Variables can be declared using any of these bases.

Decimal numbers are written as they are without any special symbols. Binary numbers are written by preceding the number with characters "0b" or "0B." Hexadecimal numbers are written by preceding the number with characters "0x" or "0X." Finally, octal numbers are written by preceding the number with number "0." The following example shows how the decimal number 55 can be represented in different bases:

55	Decimal
0b00110111	Binary
0x37	Hexadecimal
067	Octal

5.3.8 Constants

Constants are predefined variables in programs that make it easier to read a program. The following types of constants are supported:

- HIGH, LOW
- INPUT, OUTPUT
- true, false
- Integer constants
- Floating point constants

HIGH, LOW, and INPUT, OUTPUT are usually used when I/O pins are defined or when the state of a pin is set high (logic 1) or low (logic 0).

true and false are used to define 1 and 0, respectively, in our programs.

Integer and floating point constants are variables with numbers assigned to them.

5.3.9 Escape Sequences

Escape sequences are commonly used in C languages to represent nonprintable ASCII characters. For example, the character combination "\n" represents the new-line character. An escape sequence can also be represented by specifying its hexadecimal code after a backslash character. For example, the new-line character can also be represented as "\x0A." Table 5.2 gives a list of the commonly used escape sequences.

5.3.10 const Data Types

The **const** keyword is used to make a variable read-only. The value of such variables cannot be changed, and a compiler error will be generated if an attempt is made to change its value. An example is:

const float pi = 3.14159;

5.3.11 Arrays

Arrays are objects of the same type, collected under the same name. An array is declared by specifying its type, name, and the number of objects it contains. A pair of square brackets is used to specify the number of objects, starting from 0. For example:

unsigned char MyVector[5];

Table 5.2: Some commonly used escape sequences.

Escape Sequence	Hexadecimal Value	Characters
\a	0x07	BEL (bell)
\b	0x08	BS (backspace)
\f	0x0C	FF (form feed)
\n	0x0A	LF (line feed)
\r	0x0D	CR (carriage return)
\t	0x09	HT (horizontal tab)
\v	0x0B	VT (vertical tab)
\nH		String of hexadecimal digits
\\	0x5C	\ (backslash)
\'	0x27	' (single quote)
\"	0x22	" (double quote)
\?	0x3F	? (question mark)

creates an array called MyVector of type unsigned char, having five objects (or elements). The first element of the array is indexed with 0. Thus, MyVector[0] refers to the first element of the array, and MyVector[4] refers to the last element of the array. The array MyVector occupies five consecutive bytes in memory as follows:

MyVector[0]
MyVector[1]
MyVector[2]
MyVector[3]
MyVector[4]

Data can be stored in an array by specifying the array name and the index. For example, to store 50 in the second element of the above array, we have to write:

MyVector[1] = 50;   // Store 50 in the second element

Similarly, any element of an array can be copied to a variable. In the following example, the last element of the array is copied to variable **Temp**:

Temp = MyVector[4];   // Copy last element to Temp

The elements of an array can be initialised either during the declaration of the array or inside the program.

The array initialisation during the declaration is done before the beginning of the program where the values of array elements are specified inside a pair of curly brackets, and separated with commas. An example is given below where array **Months** has 12 elements and each element is initialised to the length of a month:

unsigned char Months[12] = {31, 28, 31, 30, 31, 30, 31, 31, 30, 31, 30, 31};

Thus, for example, Months[0] = 31, Months[1] = 28, and so on.

The above array could also be initialised without specifying its size. The size is filled in automatically by the compiler:

unsigned char Months[ ] = {31, 28, 31, 30, 31, 30, 31, 31, 30, 31, 30, 31};

Character arrays are declared similarly with or without specifying the size of the array. In the following example, array **Temp** stores characters of letter "COMPUTER," and the array size is set to 8 by the compiler:

unsigned char Temp[ ] = {'C', 'O', 'M', 'P', 'U', 'T', 'E', 'R'};

Strings are special cases of characters arrays that are terminated by the NULL character. The NULL character is represented by characters "\0," or by the hexadecimal number

"0x0." An example string named **MyString** is shown below that is initialised to string "COMPUTER":

unsigned char MyString[ ] = "COMPUTER";

or

unsigned char MyString[9] = "COMPUTER";

Notice that the above string declarations are automatically terminated by the compiler with a NULL character, and they are equivalent to the following character array declaration:

unsigned char MyString[ ] = { 'C', 'O', 'M', 'P', 'U', 'T', 'E', 'R', '\0' };

In the case of string declarations, the size of the array is always one more than the number of characters declared in the string.

Arrays are normally one-dimensional, and such arrays are also known as *vectors*. In C language, we can also declare arrays with *multiple dimensions*. Two-dimensional arrays are also known as *matrices*. The declaration of arrays with multiple dimensions is similar to the declaration of arrays with single dimensions. For example, a two-dimensional array is declared by specifying the array type, array name, and the size of each dimension. In the following example, **M** is declared as an integer array having the dimension 2×2 (i.e., two rows, two columns):

int M[2][2];   // A 2-dimensional array

The structure of this array is as follows, where the first element is M[0][0] and the last element is M[1][1]:

M[0][0]	M[0][1]
M[1][0]	M[1][1]

Similarly, a three-dimensional array of size $2 \times 2 \times 4$ can be declared as:

int P[2][2][4];   // A 3-dimensional array

The elements of a multidimensional array can be copied to a variable by using an assignment operator. In the following example, the data at row 1, column 2 of a two-dimensional array P is copied to variable **Temp**:

Temp = P[1][2];

Multidimensional arrays are initialised as the single-dimensional arrays where the elements are specified inside curly brackets and separated by commas. In the following example, a 3×2 integer array named **P** is declared and its elements are initialised:

int P[3][2] = { {3, 5}, {2, 1}, {1, 1} };

The structure of this array is as follows:

3	5
2	1
1	1

The above array could also be declared by omitting the first dimension (number of rows):

int P[ ][2] = { {3, 5}, {2, 1}, {1, 1} };

or

int P[ ][2] = {
        {3, 5},
        {2, 1},
        {1, 1}
    };

An interesting and useful application of multidimensional arrays is in creating two-dimensional text strings. For example, as shown below, we can create a two-dimensional text string called **wdays** to store the names of the week:

unsigned char wdays[ ] [10] = {
        "Monday",
        "Tuesday",
        "Wednesday",
        "Thursday",
        "Friday",
        "Saturday",
        "Sunday"
    };

Notice here that the first dimension (number of rows) of the array is not declared and we expect the compiler to make this dimension 7 as there are seven rows. The second dimension is set 10, because **Wednesday** is the longest string having 9 characters. If we allow 1 character for the string terminating NULL character, then a total of 10 characters will be required (each day name is terminated with a NULL character automatically).

5.3.12 Pointers

Pointers in C language are very important concepts, and they are used almost in all large and complex programs. Most students have difficulty in understanding the concept of pointers. In this section, we shall be looking at the basic principles of pointers and see how they can be used in C programs.

A pointer in C language is a variable that holds the memory address of another variable. Another simple definition of a pointer is the following: a pointer points to the memory address of a variable. Thus, by knowing where a variable is actually located in memory, we can carry out various types of operations on this variable.

Pointers in C language are declared same as any other variables, but the character "*" is inserted before the name of the variable. In general, pointers can be declared to hold the addresses of all types of variables, such as character, integer, long, and floating point, and even they can hold the addresses of functions.

The following example shows how a character pointer named **Pnt** can be declared:

> **char** *Pnt;

When a new pointer is declared, its content is not defined, and, in general, we can say that it does not point to any variable. We can assign the address of a variable to a pointer using the "&" character. In the following example, character pointer **Pnt** holds the address of (or points to the address of) character variable **Temp**:

> Pnt = &Temp;   // Pnt holds the address of Temp

Assuming variable **Temp** is at memory location 1000, pointer **Pnt** contains the number 1000, which is the address where the variable **Temp** is located at. We can now assign value to our variable **Temp** using its pointer:

> *Pnt = 25;   // Assign 25 to variable Temp

which is the same as the assignment

> Temp = 25;   // Assign 25 to Temp

Similarly, the value of **Temp** can be copied to variable **Total** as:

> Total = *Pnt;   // Copy Temp to Total

which is the same as

> Total = Temp;   // Copy Temp to Total

In C language, we can perform pointer arithmetic that may involve the following operations:

- Adding or subtracting a pointer and an integer value
- Adding two pointers
- Subtracting two pointers
- Comparing two pointers
- Comparing a pointer with a NULL

For example, assuming that the pointer **Pnt** holds address 1000, the statement

Pnt = Pnt + 1;

will increment pointer **Pnt** so that it now points to address 1001.

Pointers can be very useful when arrays are used in a program. In C language, the name of an array is also a pointer to the array. Thus, for example, consider the array:

char Sum[5];

The name **Sum** is also a pointer to this array, and it holds the address of the first element of the array. Thus, the following two statements are equal:

Sum[2] = 100;

    and

*(Sum + 2) = 100;

Since **Sum** is also a pointer to array Sum[ ], the following statement is also true:

&Sum[p] = Sum + p;

The following example shows how pointer arithmetic can be used in array-based operations. Assuming that **P** is a pointer, it can be set to hold the address of the second element of an array **Sum**:

P = Sum[2];

Suppose now that we want to clear second and third elements of this array. This can be done as:

```
*P = 0;                    // Sum[2] = 0
P = P + 1;                 // point to next location
*P = 0;                    // Sum[3] = 0;
```

or

```
*P = 0;                    // Sum[2] = 0
*(P + 1) = 0;              // Sum[3] = 0
```

or

```
Sum[2] = 0;                // Sum[2] = 0
Sum[3] = 0;                // Sum[3] = 0
```

5.3.13 Structures

Structures are similar to arrays where they store related items, but they can contain data with different data types. A structure is created using the keyword **struct**. The following example creates a structure called Family:

```
struct Family
{
        unsigned char name[20];
        unsigned char surname[30];
        unsigned char age;
        float weight;
        float height;
}
```

It is important to realise that the above is just a template for a structure and it does not occupy any space in memory. It is only when variables of the same type as the structure are created that the structure occupies space in memory. For example, variables **John** and **Peter** of type **Family** can be created by the statement:

```
struct Family John, Peter;
```

Variables of type structure can also be created during the declaration of the structure:

```
struct Family
{
        unsigned char name[20];
        unsigned char surname[30];
        unsigned char age;
        float weight;
        float height;
} John, Peter;
```

We can assign values to members of a structure by specifying the name of the structure, followed by a "." and the value to be assigned. In the following example, the age of John is set to 25 and his weight is set to 82.5:

```
John.age = 25;
John.weight = 82.5;
```

The members of a structure can also be initialised during the declaration of the structure, by specifying their values enclosed in a pair of curly brackets at the end of the structure definition, and separated with commas. In the following example, a structure called **Cube** is created with variable **MyCube** having sides A, B, and C, initialised to 2.5, 4.0, and 5.3, respectively:

```
struct Cube
{
      float sideA;
      float sideB;
      float sideC;
} MyCube = {2.5, 4.0, 5.3};
```

It is also permissible to use pointers to assign values to members of a structure. An example is given below:

```
struct Cube
{
      float sideA;
      float sideB;
      float sideC;
} *MyCube;

MyCube -> sideA = 2.5;
MyCube -> sideB = 4.0;
MyCube -> sideC = 5.3;
```

The size of a structure depends on the data types of its members. Although we can calculate the size by adding the known size of each data member, it is easier to use the operator **sizeof** to calculate the size of a structure. For example, the following statement stores the size of structure **MyCube** in variable **z**:

```
z = sizeof(MyCube);
```

Another use of structures, especially in microcontroller-based applications, is the separation of bits of a variable so that each bit or group of bits can be manipulated separately. This is also known as bit fields. With bit fields, we can assign identifiers to each bit, or to groups of bits, and then manipulate them easily. An example is given below where a byte variable called **Flag** is separated into three bit groups: the two least significant bits (bit 0 and bit 1) are called **LSB2**, the two most significant bits (bit 6 and bit 7) are called **MSB2**, and the remaining bits (bits 2 to 5) are called **Middle**:

```
struct Bits
{
      unsigned char LSB2: 2;
      unsigned char Middle: 4
      unsigned char MSB2: 2;
} Flag;
```

We can then assign values to bits of the variable:

Flag.LSB2 = 3;
Flag.Middle = 6;

5.3.14 Creating New Data Types

In C programming, we can create new data types by using the existing data types. This can be useful in many applications where we may want to give our own names to variable types. The keyword **typedef** is used to create new variable types. In the following example, a new data type called **INTEGER** is created from the data type **unsigned int**:

typedef unsigned int INTEGER;

which simply means that the new data type **INTEGER** is a synonym for the data type **unsigned int** and can be used whenever we wish to declare an **unsigned int** type variable. In the following example, **x**, **y**, and **z** are declared as **INTEGER**, which also means that they are of type **unsigned int**:

INTEGER x, y, z;

As another example, consider the following definition:

typedef char *string;

Here, string is a new data type, which is really a character pointer. We can now use this data type in our programs to create string arrays:

string Surname = "Jones";

The **typedef** keyword can also be used in structures. For example, we can create a new structure called **Person** using **typedef**:

typedef struct
{
        unsigned char name[20];
        unsigned char surname[30];
        unsigned char age;
} Person;

New variables of type Person can now be created:

Person John, Peter, Mary;

The contents of one structure can be copied to another one if both structures have been derived from the same template. An example is given below where a structure is created with two variables **Circle1** and **Circle2**. Circle1 is initialised with data and is then copied to **Circle2**:

```
struct Circle
{
        float radius;
} Circle1, Circle2;

Circle1.radius = 2.5;
Circle2 = Circle1;
```

5.3.15 Unions

Unions are similar to structures, and they are even declared and used in a similar way. The difference of a union is that all the member variables in a union share the same memory space. For example, the following union declaration creates two member variables **x** and **y**. **y** is an **unsigned integer** and occupies 2 bytes in memory. **x** is an **unsigned char** and occupies only 1 byte in memory. Because the member variables share the same memory space, the low byte of **y** shares the same memory byte as **x**:

```
union Test
{
        unsigned char x;
        unsigned int y;
}
```

If, for example, **y** is loaded with the hexadecimal data 0x2EFF, then **x** will automatically be loaded with data 0xFF.

5.3.16 Operators in C Language

Operators are used in mathematical and logical operations to produce results. Some operators are *unary* where only one operand is required, some are *binary* where two operands are required, and one operator is *tertiary* where it requires three operands.

C language supports the following operators:

- Arithmetic operators
- Relational operators
- Bitwise operators
- Logical operators
- Conditional operators
- Assignment operators
- Preprocessor operators

Table 5.3: Arithmetic operators.

Operator	Operation
+	Addition
−	Subtraction
*	Multiplication
/	Division
%	Remainder (integer division only)
++	Autoincrement
− −	Autodecrement

Arithmetic operators

Arithmetic operators are used in mathematical expressions. Table 5.3 gives a list of the arithmetic operators. All of these operators, except the autoincrement and autodecrement, require at least two operands. Autoincrement and autodecrement operators are unary as they require only one operand.

The arithmetic operators "+ − * /" are obvious and require no explanation. Operator "%" gives the remainder after an integer division is performed. Some examples are given below:

12 % 3	Gives 0 (no remainder)
12 % 5	Gives 2 (remainder is 2)
−10 % 3	Gives −1 (remainder is −1)

The autoincrement operator is used to increment the value of a variable by 1. Depending on how this operator is used, the value is incremented before (preincrement) or after (postincrement) an assignment. Some examples are given below:

```
i = 5;                    // i is equal to 5
i++;                      // i is equal to 6

i = 5;                    // i = 5
j = i++;                  // i = 6 and j = 5

i = 5;                    // i = 5
j = ++i;                  // i = 6 and j = 6
```

Similarly, the autodecrement operator is used to decrement the value of a variable by 1. Depending on how this operator is used, the value is decremented before (predecrement) or after (postdecrement) an assignment. Some examples are given below:

```
i = 5;                          // i is equal to 5
i--;                            // i is equal to 4

i = 5;                          // i = 5
j = i--;                        // i = 4 and j = 5

i = 5;                          // i = 5
j = --i;                        // i = 4 and j = 4
```

Relational operators

The relational operators are used in comparisons. All relational operators are evaluated from left to right, and they require at least two operands. Table 5.4 gives a list of the valid relational operators.

Notice that the equal operator is written with two equal signs, as "==," and not as "=." If an expression evaluates to TRUE, 1 is returned; otherwise, 0 is returned. Some examples are given below:

```
i = 5
i > 0                           // returns 1
i == 5                          // returns 1
i < 0                           // returns 0
i != 10                         // returns 1
```

Bitwise operators

Bitwise operators are used to modify bits of a variable. These operators (except the bitwise complement) require at least two operands. Table 5.5 gives a list of the bitwise operators.

Table 5.4: Relational operators.

Operator	Operation
==	Equal to
!=	Not equal to
>	Greater than
<	Less than
>=	Greater than or equal to
<=	Less than or equal to

Table 5.5: Bitwise operators.

Operator	Operation
&	Bitwise AND
|	Bitwise OR
^	Bitwise Exclusive-OR
~	Bitwise complement
<<	Shift left
>>	Shift right

Bitwise AND returns 1 if the corresponding 1 bits of the variables are 1; otherwise, it returns 0. An example is given below where the first variable is 0xFE and the second variable is 0x4F. The result of the bitwise AND operation is 0x4E:

```
0xFE:  1111 1110
0x4F:  0100 1111
       ========
       0100 1110 = 0x4E
```

Bitwise OR returns 0 if the corresponding 2 bits of the variables are 0; otherwise, it returns 1. An example is given below where the first variable is 0x1F and the second variable is 0x02. The result of the bitwise OR operation is 0x1F:

```
0x1F:  0001 1111
0x02:  0000 0010
       ========
       0001 1111 = 0x1F
```

Bitwise exclusive-OR returns 0 if the corresponding 2 bits of the variables are the same; otherwise, it returns 1. An example is given below where the first variable is 0x7F and the second variable is 0x1E. The result of the exclusive-OR operation is 0x61:

```
0x7F:  0111 1111
0x1E:  0001 1110
       ========
       0110 0001 = 0x61
```

The complement operator requires only one operand, and it complements all bits of its operand. An example is given below where variable 0x2E is complemented to give 0xD1:

```
0x2E:  0010 1110
       1101 0001 = 0xD1
```

The left shift operator requires two operands: the variable to be shifted and the count specifying the number of times the variable is to be shifted. When a variable is shifted left, 0 is filled into the least significant bit position. Shifting a variable left by one digit is same as multiplying the variable by 2. An example is given below where the variable 0x1E is shifted left by two digits to give the number 0x78:

```
0x1E:  0001 1110
       0011 1100                // shift left by one digit
       0111 1000 = 0x78         // shift left by two digits
```

The right shift operator requires two operands: the variable to be shifted and the count specifying the number of times the variable is to be shifted. When a variable is shifted right, 0 is filled into the most significant bit position. Shifting a variable right by one digit is same as dividing the variable by 2. An example is given below where the variable 0x1E is shifted right by one digit to give the number 0x0F:

```
0x1E:  0001 1110
       0000 1111 = 0x0F         // shift right by one digit
```

Logical operators

The logical operators are used in comparisons, and they return TRUE (or 1) if the comparison evaluates to nonzero or FALSE (or 0) if the comparison evaluates to zero. Table 5.6 gives a list of the logical operators.

Some examples are given below to show how the logical operators can be used:

```
a = 10;
a > 0 && a < 100              // returns 1
a > 0 || a < 20              // returns 1
a == 10 || a > 0             // returns 1
a > 0 && a > 10              // returns 0
```

Table 5.6: Logical operators.

Operator	Operation
&&	AND
||	OR
!	NOT

Conditional operator

C language supports a conditional operator with the following syntax:

result = expression1 ? expression2: expression3

Here, **expression1** is evaluated, and if its value is TRUE (nonzero), then **expression2** is assigned to **result**; otherwise, **expression3** is assigned to **result**. An example use of the conditional operator is given below which assigns the larger of variables **a** or **b** to **max**. If **a** is greater than **b**, then **a** is assigned to **max**; otherwise, **b** is assigned to **max**:

max = (a > b) ? a: b;

Another common application of the conditional operator is to convert a lower case character to upper case as shown below. In this example, if the character **c** is lower case (between "a" and "z"), then it is converted to upper case by subtracting hexadecimal 0x20 from its value:

c = (c >= 'a' && c <= 'z') ? (c − 0x20): c;

Similarly, we can write a conditional operator-based statement to convert an upper case character to lower case:

c = (c >= 'A' && c <= 'Z') ? (c + 0x20): c;

Assignment operators

The basic assignment operator is the "=" sign where a constant or an expression is assigned to a variable:

i = 200;
j = i * 2 + p;

The C language also supports compound assignment operators in the following format:

result compound operator = expression

where the compound operator can be one of the following:

```
+=     -=     *=     /=     %=
&=     |=     ^=     >>=    <<=
```

Some examples are given below to show how the compound operators can be used:

```
i += j;                    // equivalent to i = i + j
i += 1;                    // equivalent to i = i + 1
j *= p;                    // equivalent to j = j * p;
i <<= 1;                   // equivalent to i = i << 1;
```

Preprocessor operators

The preprocessor operators are an important part of the C language as these operators allow a programmer to:

- Compile a program segment conditionally
- Replace symbols with other symbols
- Insert source files to a program

A preprocessor operator is identified by the character "#," and any program line starting with this character is treated as a preprocessor operator. All preprocessor operator operations are handled by the preprocessor, just before the program is compiled. The preprocessor operators are not terminated with a semicolon.

Some of the commonly used preprocessor operators are the following:

> #define
> #undef
> #include

#define preprocessor operator is a macro expansion operator where every occurrence of an identifier in a program is replaced with the specified value of the identifier. In the following example, **MAX** is defined to be number 1000, and everywhere **MAX** is encountered in the program it will be replaced with 1000:

> **#define** MAX 1000

It is important to realise that an identifier that has been defined cannot be defined again before its definition is removed. The preprocessor operator **#undef** is used to remove the definition of an identifier. Thus, we could redefine **MAX** to be 10 as follows:

> **#undef** MAX
> **#define** MAX 10

The **#define** preprocessor operator can also be used with parameters, as in a macro expansion with parameters. These parameters are local to the operator, and it is not necessary to use the same parameter names when the operator is used (or called) in a program. An example is given below where the macro **INCREMENT** is defined to increment the value of its parameter by 1:

> **#define** INCREMENT(a) (a + 1)

When this macro is used in a program, the value of the variable representing the parameter will be incremented by 1 as in the following example:

> z = INCREMENT(x);   will change by the pre-processor to z = (x + 1);

Similarly, we can define a macro to return the larger of two numbers as:

> **#define** MAX(a, b) ((a > b) ? a: b)

And then use it in a program as:

p = MAX(x, y);

The above statement will be changed by the preprocessor to:

p = ((x > y) ? x: y);

It is important to make sure that the macro expansion is written with brackets around it in order to avoid confusion. Consider the following example that is written with no brackets:

#define ADD(a) a + 1

Now, if we use this macro in an expression as follows:

z = 2 * ADD(x);

The above statement will be expanded into z = 2 * a + 1, which is not same as z = 2 * (a + 1).

The **#include** preprocessor operator is used to include a sources file (text file) in a program. Usually program header files with extension **.h** are included in C programs:

#include <MyFile>

5.3.17 Modifying the Flow of Control

Statements in a program are normally executed in sequence one after the other until the end of the program is reached. In almost all programs, we wish to compare variables with each other, and then execute different parts of the program based on the outcome of these comparisons. We may also want to execute part of a program several times. C language provides several programming tools for modifying the normal flow of control in a program.

The following flow control statements are available in C language:

- Selection statements
- Iteration statements
- Unconditional modification of the flow of control statements

Selection statements

There are two basic selection statements: **if** and **switch**.

if statement

The **if** statement is used to carry out comparisons in a program and then force one or another part of the program to execute based on the outcome of the comparison. The basic format of the **if** statement is:

if(condition)expression;

Thus, the **expression** will be executed if the **condition** evaluates to TRUE:

> **if**(condition)
> > expression;

or

> **if**(condition)
> > expression1;
> **else**
> > expression2;

Thus, **expression1** will execute if the **condition** evaluates to TRUE; otherwise, **expression2** will be executed.

An example is given below where if **x** is equal to 5, then **z** is incremented by 1; otherwise, **z** is decremented by 1:

> **if**(x == 5)
> > z++;
> **else**
> > z--;

In many applications, we may wish to include more statements if a comparison evaluates TRUE or FALSE. This can be done by enclosing the statements in a pair of curly brackets as shown in the following example:

> **if**(x > 10 && y > 0)
> {
> > a++;
> > b = a;
> > p = 2 * b + 1;
> }

or

> **if**(x > 10 && y > 0)
> {
> > a++;
> > b = a;
> > p = 2 * b + 1;
> }
> **else**
> {
> > q = 2 * p − 2;
> > a++;
> }

The **if** statements can be nested in complex comparison operations. An example is given below:

```
if(x == 1)
        a++;
else if(x == 2)
        b++;
else if(x == 3)
        c++;
else
        d++;
```

A more practical example is given below to illustrate how the **if** statements can be nested.

Example 5.1

In an experiment, it was observed that the relationship between X and Y is found to be as follows:

X	Y
1	3.5
2	7.1
3	9.4
4	15.8

Write the **if** statements that will return Y value, given the X value.

Solution 5.1

The required program code is:

```
unsigned char x;
float Y;

if(x == 1)
        Y = 3.5;
else if(x == 2)
        Y = 7.1;
else if(x == 3)
        Y = 9.4;
else if(x == 4)
        Y = 15.8;
```

switch statement

The switch statement is the second selection statement. This statement is used when we want to compare a given variable with a different number of conditions and then execute different codes based on the outcome of each comparison. The general format of the switch statement is:

```
switch(variable)
{
        case condition1:
                statements;
                break;
        case condition2:
                statements;
                break;
        .....................
        .....................
        case condition:
                statements;
                break;
        default:
                statements;
}
```

Here, the **variable** is compared with **condition1**, and if the result evaluates to TRUE, the statements in that block are executed until the **break** keyword is encountered. This keyword forces the program to exit the switch statement by moving to the statement just after the closing curly bracket. If the **variable** is not equal to **condition1**, then it is compared with **condition2**, and if the result is TRUE, the statements in that block are executed. If none of the conditions are satisfied, then the statements under the **default** keyword are executed and then the switch statement terminates. The use of the **default** keyword is optional and can be omitted if we are sure that one of the comparisons will always evaluate to TRUE.

Perhaps the easiest way to learn how the switch statement operates is to look at some examples.

Example 5.2

Rewrite the program code in Example 5.1 using a switch statement.

Solution 5.2

The required program code is:

```
unsigned char x;
float Y;

switch (x)
{
        case 1:
                Y = 3.5;
                break;
        case 2:
                Y = 7.1;
                break;
        case 3:
                Y = 9.4;
                break;
        case 4:
                Y = 15.8;
}
```

Example 5.3

Write the program code to convert hexadecimal numbers A–F to decimal using a switch statement.

Solution 5.3

Assuming that the hexadecimal number (letter) is in variable **c**, the required program code is:

```
switch(c)
{
        case 'A':
                Y = 10;
                break;
        case 'B':
                Y = 11;
                break;
```

```
                case 'C':
                        Y = 12;
                        break;
                case 'D':
                        Y = 13;
                        break;
                case 'E':
                        Y = 14;
                        break;
                case 'F'
                        Y = 15;
                        break;
        }
```

Iteration statements

Iteration statements are used in most programs as they enable us to perform loops in our programs and repeat part of the code specified number of times. For example, we may want to calculate the sum of numbers between 1 and 10. In such a calculation, we will form a loop and keep adding the numbers from 1 to 10. C language supports the following iteration statements:

- **for** statements
- **while** statements
- **do while** statements
- **goto** statements

for statements

The for statement is perhaps the most widely used statement to create loops in programs. The general format of this statement is:

```
for(initial loop counter; conditional expression; change in loop counter)
{
        statements;
}
```

A loop variable is used in **for** loops. The initial expression specifies the starting value of the loop variable, and this variable is compared with some condition using the conditional expression before the looping starts. The statements enclosed within the pair of curly brackets are executed as long as the condition evaluates to TRUE. The value of the loop counter is changed (usually incremented by 1) at each iteration. Looping stops when the condition evaluates to FALSE.

An example is given below where a loop is formed and inside this loop the statements are executed 10 times:

```
for(i = 0; i < 10; i ++)
{
        statements;
}
```

or

```
for(i = 1; i <= 10; i++)
{
        statements;
}
```

Example 5.4

Write the program code using the **for** statement to calculate the sum of integer numbers from 1 to 10. Store the result in an integer variable called **Sum**.

Solution 5.4

The required program code is given below:

```
unsigned char i, Sum;

Sum = 0;
for(i = 1; i <= 10; i++)
{
        Sum = Sum + i;
}
```

All the parameters in a **for** loop are optional and can be omitted if required. For example, if the condition expression (middle expression) is omitted, it is assumed to be always TRUE. This creates an endless loop. An example is given below where the loop never terminates and the value of **i** is incremented by 1 at each iteration:

```
for(i = 0; ; i++)
{
        statements;
}
```

An endless loop can also be created if all the parameters are omitted as is in the following example:

```
for(;;)
{
        statements;
}
```

Endless loops are used frequently in microcontroller programs. An example of creating an endless loop using a preprocessor command and a **for** statement is given in the following example:

Example 5.5

Write the program code to create an endless loop using a **for** statement. Create a macro called **DO_FOREVER** to implement the loop.

Solution 5.5

The required program code is given below:

```
#define DO_FOREVER for(;;)
………………..
………………..
DO_FOREVER
{
        statements;
}
```

If we omit the second and third parameters of a **for** loop, then the loop repeats endless as in the earlier example, but the value of the initial loop counter never changes. An example is given below where the value of **i** is always 0:

```
for(i = 0;;)
{
        statements;
}
```

for loops can be nested such that in a two-level loop the inner loop is executed for each iteration of the outer loop. Two-level nested loops are commonly used in matrix operations, such as adding two matrices and so on. A two-level nested **for** loop is given below. In this example, the statements in the inner loop are executed 100 times:

```
for(i = 0; i < 10; i++)
{
        for(j = 0; j < 10; j++)
        {
                statements in inner loop;
        }
}
```

An example nested loop to add all the elements of a 2 × 2 matrix **M** is given below.

Example 5.6

Write the program code using **for** loops to add elements of a 2 × 2 matrix called **M**.

Solution 5.6

The required program code is given below:

```
Sum = 0;
for(i = 0; i < 2; i++)
{
        for(j = 0; j < 2; j++)
        {
                Sum = Sum + M[i][j];
        }
}
```

while statement

The **while** statement is another statement that can be used to create loops in programs. The format of this statement is:

```
while (condition)
{
        statements;
}
```

Here, the loop enclosed by the curly brackets is executed as long as the specified **condition** evaluates to TRUE. Notice that if the **condition** is FALSE on entry to the loop, then the loop never executes. An example is given below where the statements inside the **while** loop execute 10 times:

```
i = 0;
while (i < 10)
{
        statements;
        i++;
}
```

Here, at the beginning of the loop **i** is equal to 0, which is less than 10 and therefore the loop starts. Inside the loop, **i** is incremented by 1. The loop terminates when **i** becomes equal to 10. It is important to realise that the condition specified at the beginning of the loop should be satisfied inside the loop; otherwise, an endless loop is formed.

In the following example, **i** is initialised to 0, but is never changed inside the loop and is therefore always less than 10. Thus, this loop will never terminate:

```
i = 0;
while (i < 10)
{
        statements;
}
```

It is possible to have a **while** loop with no statements. Such a loop is commonly used in microcontroller applications to wait for certain action to be completed, for example, to wait until a button is pressed. An example is given below where the program waits until variable CLK becomes 0:

```
while(CLK == 1);
```

or

```
while(CLK);
```

while loops can be nested inside one another. An example is given below:

```
i = 0;
j = 0;
while (i < 10)
{
        while (j < 10)
        {
                statements;
                j++;
        }
        statements;
        i++;
}
```

do while statement

The **do while** statement is similar to the while statement, but here the condition for the existence of the loop is at the end of the loop. The format of the **do while** statement is:

```
do
{
        statements;
} while (condition);
```

The loop executes as long as the **condition** is TRUE, that is, until the **condition** becomes FALSE. Another important difference of the **do while** loop is that because the condition is at the end of the loop, the loop is always executed at least once. An example of the **do while** loop is given below where the statements inside the loop are executed 10 times:

```
i = 0;
do
{
        statements;
        i++;
} while (i < 10);
```

The loop starts with i = 0, and the value of **i** is incremented inside the loop. The loop repeats as long as **i** is less than 10, and it terminates when **i** becomes equal to 10.

As with the **while** statement, an endless loop is formed if the condition is not satisfied inside the loop. For example, the following **do while** loop never terminates:

```
i = 0;
do
{
        statements;
} while (i < 10);
```

goto statement

The **goto** statement can be used to alter the flow of control in a program. Although the **goto** statement can be used to create loops with finite repetition times, use of other loop structures such as **for, while,** and **do while** is recommended. The use of the **goto** statement requires a label to be defined in the program. The format of the **goto** statement is:

```
        ..............
Label:
        ..............
        goto Label;
```

The label can be any valid variable name, and it must be terminated with a colon character. The following example shows how the **goto** statement can be used in a program to create a loop to iterate 10 times:

```
        i = 0;
Loop:
        statements;
        i++;
        if(i < 10) goto Loop;
        …………………..
        …………………..
```

Here, label **Loop** identifies the beginning of the loop. Initially i = 0 and is incremented by 1 inside the loop. The loop continues as long as i is less than 10. When i becomes equal to 10, program terminates and continues with the statement just after the **goto** statement.

Unconditional modification of flow of control

As described above, the **goto** statement is used to unconditionally modify the flow of control in a program. Although use of the **goto** statement is not recommended as it may give rise to unreadable and unmaintainable code, its general format is as follows:

```
        …………………..
Label:
        …………………..
        …………………..
        goto Label;
```

Creating infinite loops

As discussed earlier in this chapter, infinite loops are commonly used in microcontroller-based applications. Such loops can be created using all of the flow control statement we have seen. Table 5.7 summarises the ways that an infinite loop can be created.

Table 5.7: Methods of creating an infinite loop.

for Loop	while Loop	do while Loop	goto Loop

for(;;)
{
statements;
}
	while
	(1)
	{
	statements;
	}
		do
		{
		statements;
		}
		while(1);
			Loop:
			statements;
			goto
			Loop;

Premature termination of a loop

There are applications where we may want to terminate a loop prematurely, that is, before the loop-terminating condition is satisfied. The **break** statement can be used for this purpose. An example is given below where the **for** loop is terminated when the loop counter **i** becomes equal to 10:

```
for(i = 0; i < 100; i++)
{
        statements;
        if(i == 10)break;
}
```

Another example is given below where the **while** statement is used to create the loop:

```
while (i < 10)
{
        statements;
        if(i == 3)break;
}
```

Skipping an iteration

The **continue** statement is used to skip an iteration inside a loop. This statement, rather than terminating a loop, forces a new iteration of the loop to begin, by skipping any statements left in that iteration. An example is given below using the **for** loop where iteration 3 is skipped:

```
for(i = 0; i < 10; i++)
{
        if(i == 3)continue;                          // skip iteration 3
        statements;
}
```

5.4 Functions

It is widely accepted that a large program should, wherever possible, be expressed in terms of a number of smaller, separate functions. Each function should be a self-contained section of the code written to perform a specific task. Such an approach makes the programming, and more importantly the testing and maintenance of the overall program, much easier. For example, during program development, each function can be developed and tested independently. When the individual functions are tested and working, they can be included in the overall program. The main program can be a few lines long, and can consist of function calls to carry out the required operations.

In general, most functions perform some tasks and then return data to the calling program. It is not however a requirement that a function must return data. Some functions do only certain tasks and do not return any data. For example, a function to turn on an LED is not expected to return any data.

5.4.1 Functions in C Language

In C language, the format of a function is as follows:

```
type name (parameter1, parameter2,.....)
{
        function body
}
```

At the beginning of a function declaration, we first declare the data type to be returned by the function. If the function is not expected to return any data, the keyword void should be inserted before the function name. Then, we specify the unique name of the function. A function name can be any valid C variable name. This is followed by a set of brackets where function parameters are specified, separated by commas. Some functions may not have any parameters, and empty brackets should be used while declaring such functions. The body of the function contains the code to be executed by the function, and this code must be enclosed within a pair of curly brackets.

An example function declaration with the name **INC** is shown below. This function receives a parameter of type **int**, multiplies the value of this parameter by 2, and then returns the result to the calling program:

```
int INC(int a)
{
        return (2 * a);
}
```

Notice that the keyword **return** must be used if it is required to return value to the calling program. The data type of the returned number must match with the data type specified before the function name.

The above function can be called from a program as follows:

```
z = INC(x);
```

which will increment the value of variable **x** by 1. Notice that the number of parameters supplied by the calling statement must match with the number of parameters the function expects. Also, the variables used in a function are local to the function and do not have any connection with variables with the same name in the main program. As a result of this, variable names used in the parameters do not have to be same as the names used while calling the function.

If a function is not expected to return any value, then the **return** statement can be omitted. In such an application, the function will automatically return to the calling program when it terminates. An example function with no return data is given below. This function simply sets LED1 to 1:

```
void LED()
{
        LED1 = 1;
}
```

void functions are called with no parameters:

```
LED();
```

Although functions are usually declared before the main program, this is not always the case as we shall see later.

Some example programs are given below to illustrate how functions may be used in main programs.

Example 5.7

Write a program to calculate the area of a circle whose radius is 3.5 cm. The program should call a function named **Area_Of_Circle** to calculate and return the area. The radius of the circle should be passed as a parameter to the function.

Solution 5.7

The area of a circle with radius **r** is given by:

$$\text{Area} = \pi r^2$$

Since the radius and the area are floating point numbers, the function type and the parameter must be declared as **float**.

The required function is shown in Figure 5.5. This function is used in a main program as shown in Figure 5.6. Notice that comments are used throughout the program to clarify the operation of the program. The function is called with the parameter set to 5.6 to calculate the

```
float Area_Of_Circle(float radius)
{
        float area;

        area = 3.14 * radius * radius;
        return (area);
}
```

Figure 5.5: Function to Calculate the Area of a Circle

```
/***********************************************************

                    AREA OF A CIRCLE
                    ===============

This program calculates the area of a circle with a radius of 3.5cm.


    Programmer:    Dogan Ibrahim
    Date:          May, 2014
    File:          Circle.C
***********************************************************/

//
// This function calculates and returns the area of a circle, given its radius
//
float Area_Of_Circle(float radius)
{
        float area;

        area = 3.14 * radius * radius;
        return (area);
}

void setup()
{
}

//
// Start of MAIN program
//
void loop()
{
        float r, a;

        r = 3.5;
        a = Area_Of_Circle(r);

        while(1);
}
```

Figure 5.6: Program to Calculate the Area of a Circle

area of the circle. Notice that an endless loop is formed at the end of the program using the **while(1)** statement. This is because it is not required to repeat the main program.

Example 5.8

Write a program to calculate the area and circumference of a rectangle whose sides are 2.4 and 6.5 cm. The program should call functions named **Area_Of_Rectangle** and **Circ_Of_Rectangle** to calculate and return the area and the circumference of the rectangle, respectively.

Solution 5.8

The circumference and the area of a rectangle with sides a and b are given by:

$$\text{Circumference} = 2 \times (a + b)$$

$$\text{Area} = a \times b$$

Since the inputs and outputs are floating point numbers, we shall be declaring the variable as float.

The required functions are shown in Figure 5.7. These functions are used in a main program as shown in Figure 5.8. Notice that comments are used throughout the program to clarify the operation of the program. The function is called with the sides of the rectangle set to 2.4 and 6.5 cm.

```
float Circ_Of_Rectangle(float a, float b)
{
        float circumference;

        circumference = 2 * (a + b);
        return (circumference);
}


float Area_Of_Rectangle(float a, float b)
{
        float area;

        area = a * b;
        return (area);
}
```

Figure 5.7: Function to Calculate the Circumference and Area of a Rectangle

```
/*******************************************************************

                    CIRCUMFERENCE AND AREA OF A RECTANGLE
                    ========================================

This program calculates the circumference and area of a rectangle with sides
2.4cm and 6.5cm

Programmer:   Dogan Ibrahim
Date:         May, 2014
File:         Rectangle.C
*******************************************************************/
//
// This function calculates and returns the circumference of a rectangle, given its sides
//
float Circ_Of_Rectangle(float a, float b)
{
        float circumference;

        circumference = 2 * (a + b);
        return (circumference);
}


//
// This function calculates and returns the area of a rectangle, given its sides
//
float Area_Of_Rectangle(float a, float b)
{
        float area;

        area = a * b;
        return (area);
}
// This function calculates and returns the area of a circle, given its radius


void setup()
{
}


//
// Start of MAIN program
//
void loop()
{
        float a = 2.4, b = 6.5;
        float circ, area;

        circ = Circ_Of_Rectangle(a, b);
        area = Area_Of_Rectangle(a, b);

        while(1);
}
```

Figure 5.8: Program to Calculate the Circumference and Area of a Rectangle

Example 5.9

Write a function called **HEX_CONV** to convert a single-digit hexadecimal number "A" to "F" to decimal integer. Use this function in a program to convert hexadecimal "E" to decimal.

Solution 5.9

The required function code is shown in Figure 5.9. The hexadecimal number to be converted is passed as a parameter to the function. A switch statement is used in the function to convert and return the given single-digit hexadecimal number to decimal. The main program listing is given in Figure 5.10. Notice that hexadecimal number to be converted is passed as a parameter when calling the function **HEX_CONV**.

```
unsigned char HEX_CONV(unsigned char c)
{
        unsigned char y;

        switch(c)
        {
                case 'A':
                        y = 10;
                        break;
                case 'B':
                        y = 11;
                        break;
                case 'C':
                        y = 12;
                        break;
                case 'D':
                        y = 13;
                        break;
                case 'E':
                        y = 14;
                        break;
                case 'F'
                        y = 15;
                        break;
        }
        return (y);
}
```

Figure 5.9: Function to Convert Hexadecimal to Decimal

```
/*******************************************************************

              HEXADECIMAL TO DECIMAL CONVERSION
              ===================================

This program converts the single digit hexadecimal number "E" to decimal.
Function HEX_CONV is used to convert a given single digit hexadecimal number
to decimal.

Programmer:    Dogan Ibrahim
Date:          May, 2014
File:          Hex.C
*******************************************************************/
//
// This function converts a given single digit hexadecimal number A-F to decimal
//
unsigned char HEX_CONV(unsigned char c)
{
        unsigned char y;

        switch(c)
        {
                case 'A':
                        y = 10;
                        break;
                case 'B':
                        y = 11;
                        break;
                case 'C':
                        y = 12;
                        break;
                case 'D':
                        y = 13;
                        break;
                case 'E':
                        y = 14;
                        break;
                case 'F':
                        y = 15;
                        break;
        }
        return (y);
}


void setup()
{
}

//
// Start of MAIN program
//
void loop()
{
        unsigned char hex_number = 'E';
        unsigned char r;

        r = HEX_CONV(hex_number);

        while(1);
}
```

Figure 5.10: Program to Convert Hexadecimal Number "E" to Decimal

5.4.2 Passing Parameters to Functions

Functions are the building blocks of programs where we divide a large program into manageable smaller independent functions. As we have seen earlier, functions generally have parameters supplied by the calling program. There are many applications where we may want to pass arrays to functions and let the function carry out the required array manipulation operations.

Passing single elements of an array to a function is simple, and all we have to do is specify the required elements when calling the function. In the following statement, the second and third elements of array **A** are passed to a function called **Sum**, and the result returned by the function is assigned to variable **b**:

   b = Sum(A[1], A[2]);

Passing an array by reference

In most array-based operations, we may want to pass an entire array to a function. This is normally done by specifying the address of the first element of the array in the calling program. Since the address of the first element of an array is same as the array name itself, we can simply specify the name of the array. In the function declaration, we have to specify the array name with a set of brackets where the size of the array is not specified. This way, all the elements of the array will be available to the function. Notice that here we are passing the entire array by **reference**, which means that the original array elements can be modified inside the function.

The following statements show how the entire array called **A** is passed to a function called **Sum**:

   In the calling program: Sum(A);

   In the function header: type Sum(A[ ])
                                    {
                                            body of the function
                                    }

Some complete examples are given below.

Example 5.10

Write a program to store numbers 1–10 in an integer array called **A**. Then, call a function to calculate and return the sum of all the array elements.

Solution 5.10

The required program listing is shown in Figure 5.11. In the main program, a **for** loop is used to store integer numbers 1–10 in an array called **A**. Then, function **Sum** is called to

```
/********************************************************************

                PASSING AN ARRAY TO A FUNCTION BY EFERENCE
                ===========================================

This program shows how an entire array can be passed to a function by reference.
In this example, the name of the array (i.e. the address of the first array element) is
passed as a parameter to the array.

The array calculates and returns the sum of all the array elements.


Programmer:    Dogan Ibrahim
Date:          May, 2014
File:          Array1.C
********************************************************************/
//
// This function calculates and returns the sum of all the array elements
//
unsigned char Sum(unsigned char Nums[ ])
{
        unsigned char i, Sum;

        Sum = 0;
        for(i= 0; i < 10; i++)Sum = Sum + Nums[i];
        return (Sum);
}


void setup()
{
}

//
// Start of MAIN program
//
void loop()
{
        unsigned char  A[10];
        unsigned char i, Sum_Of_Numbers;

        for(i= 0; i < 10; i++)A[i] = i;
        Sum_Of_Numbers = Sum(A);

        while(1);
}
```

Figure 5.11: Program Passing an Array by Reference

calculate and return the sum of all the array elements. The returned value is stored in variable **Sum_Of_Numbers**. The array is passed to the function by specifying its starting address, that is, its name.

Example 5.11

Repeat Example 5.10, but this time define the array size at the beginning of the program using the **#define** preprocessor operator and then pass both the array and its size to the function.

Solution 5.11

The required program listing is shown in Figure 5.12. Here, the array size is defined to be 50. Notice also that the data types are changed to **unsigned integer** as the sum could be larger than 255.

Passing an array using pointers

It is also possible to pass an entire array to a function using pointers. In this method, the address of the array (i.e., its name) is specified in the calling program. In the function header, a pointer declaration is used and the elements of the array are manipulated using pointer operations as we have seen earlier. An example is given below to illustrate the concept.

Example 5.12

Repeat Example 5.11, but this time pass the array to the function using pointers.

Solution 5.12

The required program listing is given in Figure 5.13. Initially, array **A** is loaded with integer numbers 0–10. Then, function **Sum** is called to calculate and return the sum of all array elements. The address of array **A** is passed in the calling program by specifying its name. In the function header, a pointer declaration is made so that the elements of the array can be accessed in the function body using pointer operations.

Passing a string using pointers

As we have seen before, a string is a character array, terminated with a NULL character. An array can be passed to a function using pointers as in the earlier example. An example is given below to illustrate the concept.

Example 5.13

Write a program to declare a string with the text **Learning to program in C** in the main program. Then, call a function to calculate the number of times character **a** is used in the string and return this number to the main program.

```
/******************************************************************

                 PASSING AN ARRAY TO A FUNCTION BY REFERENCE
                 ==============================================
```

This program shows how an entire array can be passed to a function by reference. In this example, the name of the array (i.e. the address of the first array element) is passed as a parameter to the array.

In addition to passing the entire array, the array size is also passed to the function. The array calculates and returns the sum of all the array elements.

```
Programmer:    Dogan Ibrahim
Date:          May, 2014
File:          Array2.C
******************************************************************/
#define Array_Size 50

//
// This function calculates and returns the sum of all the array elements
//
unsigned int Sum(unsigned int Nums[ ], unsigned int N)
{
        unsigned int i, Sum;

        Sum = 0;
        for(i= 0; i < N; i++)Sum = Sum + Nums[i];
        return (Sum);
}


void setup()
{
}

//
// Start of MAIN program
//
void loop()
{
        unsigned int  A[Array_Size];
        unsigned int i, Sum_Of_Numbers;

        for(i= 0; i < Array_Size; i++)A[i] = i;
        Sum_Of_Numbers = Sum(A, Array_Size);

        while(1);
}
```

Figure 5.12: Another Program Passing an Array by Reference

```
/*********************************************************************

                PASSING AN ARRAY TO A FUNCTION USING POINTERS
                ===============================================
```

This program shows how an entire array can be passed to a function by using pointers. In this example, the name of the array (i.e. the address of the first array element) is passed as a parameter to the array.

The array calculates and returns the sum of all 10 array elements.


```
Programmer:     Dogan Ibrahim
Date:           May, 2014
File:           Array3.C
*********************************************************************/
//
// This function calculates and returns the sum of all the array elements
//
unsigned char Sum(unsigned char *Nums)
{
        unsigned char i, Sum;

        Sum = 0;
        for(i= 0; i < 10; i++)Sum = Sum + *(Nums + i);
        return (Sum);
}


void setup()
{
}

//
// Start of MAIN program
//
void loop()
{
        unsigned char  A[10];
        unsigned char i, Sum_Of_Numbers;

        for(i= 0; i < 10; i++)A[i] = i;
        Sum_Of_Numbers = Sum(A);

        while(1);
}
```

Figure 5.13: Program Passing an Array Using Pointers

Solution 5.13

The required program listing is given in Figure 5.14. The text is stored in a string called **Txt**. This string and the character to be matched are passed to a function called **COUNT**. Pointer operations are used inside the function to find out how many times the required character is used in the string. This count is returned to the main program and stored in variable **Sum**. Notice that in the main program the string is declared using a character array where the size of the array is automatically initialised by the compiler:

   unsigned char Txt[ ] = "Learning to program in C";

We could also have declared the same string using a pointer as:

   unsigned char *Txt = "Learning to program in C";

5.4.3 Passing Variables by Reference to Functions

In C language, when we call a function, the variables in function parameters are passed **by value** by default. This means that only the values of the variables are passed to the function and these values cannot be modified inside the function. There are applications, however, where we may want to modify the values of the variables passed as parameters. In such applications, we have to pass the parameters **by reference**. Passing a parameter **by reference** requires the address of the parameter to be passed to the function so that the value of the parameter can be modified inside the function. It is important to realise that such modifications are permanent and affect the actual variable in the main program that is used while calling the function. In order to pass parameters **by reference**, we have to use pointers. An example is given below to illustrate the concept of how a parameter can be passed to a function **by reference**.

Example 5.14

Write a function to receive an integer number and then increment its value by 1. Use this function in a program and pass the value to be incremented to the function **by reference**.

Solution 5.14

The required function listing is given in Figure 5.15. In the main program, the variable **Total** is initialised to 10. Function **INCR** is then called, and the address of **Total** is passed to the function. The function increments **Total** by 1. Notice that the function type is declared as **void** since it does not return any value to the calling program.

```
/********************************************************************

                 PASSING A STRING USING POINTERS
                 ===============================

This program shows how a string can be passed to a function by using pointers. In this
example, a string is declared in the main program and then a function is called to find
out how many times the character "a" is used in the string. This count is returned to the
main program.


Programmer:   Dogan Ibrahim
Date:         May, 2014
File:         Array4.C
********************************************************************/
//
// This function calculates how many times the character c is used in the string and
// returns this count to the calling program
//
unsigned char Count(unsigned char *str, unsigned char c)
{
        unsigned char Cnt = 0;

        while(*str)                             // do while not a NULL character
        {
                if(*str == c)Cnt++;             // if  a matching character
                str++;                          // point to next character in string
        }
        return (Cnt);                           // return the count
}


void setup()
{
}

//
// Start of MAIN program
//
void loop()
{
        unsigned char  Txt[ ] = "Learning to program in C";
        unsigned  char Sum;

        Sum = Count(Txt, 'a');

        while(1);
}
```

Figure 5.14: Passing a String to a Function

```
/**********************************************************************

                PASSING A VARIABLE TO A FUNCTION BY REFERENCE
                ================================================
```

This program shows how a variable can be passed to a function by reference. The
address of the variable is passed to the function. In this example the function
increments the value of the variable by one. It is important to realize that since the
address of the variable is passed the change to the variable is permanent. i.e. the
value of the actual variable used in the calling program is modified by the function.


```
Programmer:    Dogan Ibrahim
Date:          May, 2014
File:          Byref.C
**********************************************************************/
//
// This function increment the value of its parameter by one. The parameter is passed
// by reference. i.e. the actual address of the variable is passed to the function. Notice
// the function is declared as void as it does not return any value to the calling program
//
void INCR(int *a)
{
        *a = *a + 1;
}


void setup()
{
}

//
// Start of MAIN program
//
void loop()
{
        int Total;

        Total = 10;                                 // initialize p to 10
        INCR(&Total);                               // Increment value of Total by one

        while(1);
}
```

Figure 5.15: Passing a Variable by Reference

5.4.4 Static Variables in Functions

Normally, the variables used in functions are local to the function and they are re-initialised
(usually to 0) when the function is called. There are some applications, however, where we
may want to preserve the values of some or all of the variables inside a function declaration.

This is done using the static statement. An example function code is given below. Here, variable **Total** is initialised to 0 using the static keyword. Notice that the variable is incremented by 1 before returning from the function. What happens now is that every time the function is called, variable **Total** will retain its previous value. Thus, on the second call, **Total** will be 2, on the third call it will be 3, and so on:

```
void Sum(void)
{
        static Total = 0;
        ………………..
        ………………..
        Total++;
}
```

5.4.5 Function Prototypes

If a function is defined after a main program, then it is not visible to the compiler at the time the function is called first time in the main program. The same will happen if there are several functions in a program and a function makes call to another function that is defined at a later point in the program. In order to avoid compilation errors arising because of these compiler visibility issues, we construct function prototypes and declare them at the beginning of the program. A function prototype simply includes a function name and a list of the data types of its parameters. Function prototypes must be terminated with a semicolon. A typical function prototype statement is given below where the function is named **Test** and it has two integer and one floating point parameters:

```
int Test(int, int, float);
```

An example is given below to illustrate how function prototypes can be used in programs.

Example 5.15

Write a function to add two integer numbers and return the result to a calling program. Declare the function after the main program, and use function prototype declaration at the beginning of the program.

Solution 5.15

The required program listing is shown in Figure 5.16. Notice that the function prototype for function **Add** is declared at the beginning of the program and the function declaration is after the main program is declared.

```
/**********************************************************************

                        USING FUNCTION PROTOTYPES
                        =============================

This program shows how we can use a function prototype in a program. In this program
A function is created which adds two integer numbers.


Programmer:    Dogan Ibrahim
Date:          May, 2014
File:          Funcproto.C
**********************************************************************/
int Add(int, int);                                      // Declare function prototype

void setup()
{
}

//
// Start of MAIN program
//
void loop()
{
        int x, y, Total;

        x = 5;
        y = 8;
        Total = Add(x, y);
}


//
// This function adds two integer numbers and returns the result.
//
int Add(int a, int b)
{
        return (a + b);                                 // return the sum
}
```

Figure 5.16: Declaring a Function Prototype

5.5 Microcontroller-Specific Features

It is important to know some of the microcontroller-specific features of the C language and the chipKIT MX3 development board before developing programs and projects.

5.5.1 Microcontroller Digital Inputs and Outputs

The chipKIT MX3 development board provides 40 general-purpose I/O pins via its Pmod connectors. In addition, two I²C bus pins are provided that are not available as general-purpose I/O pins. All of the I/O pins can be configured as inputs or outputs (analogue pins can only be analogue inputs or digital inputs/outputs).

Each PIC32 microcontroller port is 16-bits wide, and they are numbered A–G (depending on the microcontroller type, some ports may be absent).

The I/O ports are controlled using four registers: TRIS, LAT, PORT, and ODC. These registers are referenced by adding the port name letter to their end. For example, PORTA has registers TRISA, LATA, PORTA, and ODCA.

The TRIS register is used to set the I/O direction of a port pin. Setting a TRIS bit to 0 makes the corresponding port pin an output. Similarly, setting a TRIS bit to 1 makes the corresponding port pin an input.

The LAT register is used to write to a port pin. Reading from this register returns the last value written to the port pin (which may not necessarily be the current state of the port pin).

The PORT register is similar to the LAT register, and writing to this register sets the corresponding port pin as requested. Reading from the PORT register returns the current state of the port.

The ODC register is used to control the output type of a port pin. Setting this register to 0 makes the corresponding port pin a normal output pin (the default). Setting to 1 makes the port pin an open-drain output.

5.5.2 Logical I/O Port Pin Numbers

The MPIDE uses logical pin numbers to identify the I/O ports. These numbers start from 0 and are numbered up consecutively. On the chipKIT MX3, the 40 I/O port pins are numbered as 0–39 on the Pmod connectors. Pmod connector JA pin 1 (JA-01) is assigned I/O pin 0, JA-02 is pin 1, and so on. Table 5.8 shows the logical port pin numbers and the corresponding physical port pin names, organised in logical pin number order. Table 5.9 shows the logical and corresponding physical pin numbers, organised in port name order. Logical port pins 40 and 41 are used for the I²C bus.

Table 5.8: Logical and physical port pin names (in logical pin number order).

Logical Pin	Physical Pin	Logical Pin	Physical Pin
0	RE00 (JA-01)	21	RB01 (JC-08)
1	RE01 (JA-02)	22	RD00 (JC-09)
2	RE02 (Ja-03)	23	RD01 (JC-10)
3	RE03 (JA-04)	24	RB02 (JD-01)
4	RE04 (JA-07)	25	RD02 (JD-02)
5	RE05 (JA-08)	26	RD10 (JD-03)
6	RE06 (JA-09)	27	RB09 (JD-04)
7	RE07 (JA-10)	28	RB12 (JD-07)
8	RD09 (JB-01)	29	RD03 (JD-08)
9	RF03 (JB-02)	30	RD11 (JD-09)
10	RF02 (JB-03)	31	RB13 (JD-10)
11	RF06 (JB-04)	32	RG09 (JE-01)
12	RD06 (JB-07)	33	RG08 (JE-02)
13	RD05 (JB-08)	34	RG07 (JE-03)
14	RD04 (JB-09)	35	RG06 (JE-04)
15	RD07 (JB-10)	36	RD08 (JE-07)
16	RB08 (JC-01)	37	RB05 (JE-08)
17	RF05 (JC-02)	38	RB04 (JE-09)
18	RF04 (JC-03)	39	RB03 (JE-10)
19	RB14 (JC-04)	42	RF00 (LD4)
20	RB00 (JC-07)	43	RF01 (LD5)

5.5.3 Controlling the I/O Ports Using MPIDE

Three built-in functions are available for controlling an I/O port when in digital mode: pin-Mode(), digitalRead(), and digitalWrite().

pinMode() function is used to set the I/O port pin direction, and it can take the values INPUT, OUTPUT, or OPEN. OPEN is similar to OUTPUT and is used for open-drain outputs.

digitalRead() function is used to read data from an I/O port. An I/O port can either be HIGH (logic 1) or LOW (logic 0).

digitalWrite() is used to set the state of an I/O port.

5.5.4 I/O Pin Voltage Levels

The PIC32 microcontroller family operates with 3.3 V supply, and the output voltage levels are 0–3.3 V. There are some applications where we may want to use 5 V signals. In general, the digital I/O pins of the PIC32 microcontroller are tolerant to 5 V inputs and we can apply 5 V signals directly to these inputs without damaging the microcontroller. The analogue inputs, on the other hand, are not 5 V-tolerant, and the maximum voltage that can be applied

Table 5.9: Logical and physical port pin names (in port name order).

Port Pins	Logical Number	Port Pins	Logical Number
RB00 (AN0)	20 (JC-07)	RD06	12 (JB-07)
RB01 (AN1)	21 (JC-08)	RD07	15 (JB-10)
RB02 (AN2)	24 (JD-01)	RD08	36 (JE-07)
RB03 (AN3)	39 (JE-10)	RD09	8 (JB-01)
RB04 (AN4)	38 (JE-09)	RD10	26 (JD-03)
RB05 (AN5)	37 (JE-08)	RD11	30 (JD-09)
RB06 (AN6)	N/A (AN8)	RE00	0 (JA-01)
RB07 (AN7)	N/A (AN7)	RE01	1 (JA-02)
RB08 (AN8)	16 (JC-01)	RE02	2 (JA-03)
RB09 (AN9)	27 (JD-04)	RE03	3 (JA-04)
RB10 (AN10)	N/A	RE04	4 (JA-07)
RB11 (AN11)	N/A	RE05	5 (JA-08)
RB12 (AN12)	28 (JD-07)	RE06	6 (JA-09)
RB13 (AN13)	31 (JD-10)	RE07	7 (JA-10)
RB14 (AN14)	19 (JC-04)	RF00	N/A (LD4)
RB15 (AN15)	N/A	RF01	N/A (LD5)
RC12	N/A	RF02	10 (JB-03)
RC13	N/A	RF03	9 (JB-02)
RC14	N/A	RF04	18 (JC-03)
RC15	N/A	RF05	17 (JC-02)
RD00	22 (JC-09)	RF06	11 (JB-04)
RD01	23 (JC-10)	RG06	35 (JE-04)
RD02	25 (JD-02)	RG07	34 (JE-03)
RD03	29 (JD-08)	RG08	33 (JE-02)
RD04	14 (JB-09)	RG09	32 (JE-01)
RD05	13 (JB-08)		

to these inputs is 3.3 V. In general, applying 5 V input to an input that is not tolerant to 5 V may damage the microcontroller circuitry. However, there are some techniques that can be used to overcome this problem. One simple technique is to use resistors (e.g., 200 Ω) in series and Schottky diodes in parallel with the input pins in order to limit the input voltages to safe levels.

The output high voltage level of the PIC32 microcontroller is typically 2.4 V, and this can be recognised as valid logic 1 by most logic families. If this voltage is not sufficient to cause a logic 1, a pull-up resistor (e.g., 5k) to 5 V can be used at the outputs of the pins that are tolerant to 5 V to increase the output voltage levels.

5.6 Built-In Functions

The C compiler contains a number of useful functions that can be used to simplify our programs. In this section, we shall be looking at some of these functions.

5.6.1 Data Conversion Functions

The following functions are available to convert one data type to another one:

char(): Converts into character data type
byte(): Converts into byte data type
int(): Converts into integer data type
word(): Converts into word data type
long(): Converts into long data type
float(): Converts into float data type

5.6.2 Digital I/O Functions

The following digital I/O functions are available. These functions are usually used in the **setup**() function at the beginning of a program:

pinMode(pin, mode): Used to set the direction of an I/O pin. Valid **modes** are INPUT and OUTPUT. An example is given below:

```
int led = 42;
pinMode(led, OUTPUT);
```

digitalWrite(pin, value): Used to set the output state of an output pin to LOW (logic 0) or HIGH (logic 1). The I/O pin must be configured as an output pin before this function is called. The **value** can be HIGH or LOW. An example is given below:

```
digitalWrite(led, HIGH);
```

digitalRead(pin): Used to read the state of an I/O pin. The I/O pin must be configured as an input pin before this function is called. The function returns HIGH or LOW. An example is given below:

```
stat = digitalRead(led);
```

5.6.3 Analogue I/O Functions

The following analogue I/O functions are available:

analogRead(pin): Used to read value from the specified analogue input pin. The function returns an integer between 0 and 1023 (10-bit A/D conversion is done). An example is given below:

```
int anapin = 5;
int anadata;
anadata = analogRead(anapin);
```

analogWrite(pin, duty-cycle): Used to send an analogue value in the form of a pulse-width-modulated (PWM) waveform to the specified pin. When this function is called, a square wave signal with a frequency of approximately 490 Hz is sent out of the pin with the specified duty cycle. The duty cycle is an unsigned character that can have values between 0 and 255. The PWM waveform is cancelled when a new operation is performed on the specified pin (e.g., digital read or write, or a new analogue write). An example is given below:

```
led = 22;
pwm = 120;
analogWrite(led, pwm);
```

This function is only available for the following I/O pins:

Pmod Connector	Logical Pin Number	Microcontroller I/O Pin
JC-09	22	RD0
JC-10	23	RD1
JD-02	25	RD2
JD-08	29	RD3
JB-09	14	RD4

analogReference(mode): Used to set the reference voltage for the analogue input. The default value is +3.3 V.

5.6.4 Other I/O Functions

The other suitable I/O functions are as follows:

tone(pin, frequency, duration): Generates a square wave signal of the specified frequency (in hertz), with 50% duty cycle. The duration (in milliseconds) is optional, and if not specified the signal will continue until a call to **noTone()** is made. In the example below, a signal with a frequency of 1000 Hz is generated on pin 12 that lasts for 500 ms:

```
tone(12, 1000, 500);
```

noTone(pin): Used to stop an active tone on a pin.

shiftOut(dataPin, clockPin, bitOrder, value): This function is used to shift out a byte (value) of data from the specified pin. The shift operation can start either from the most significant bit or from the least significant bit. The clockPin is toggled once the dataPin has been set to the correct value. The bitOrder can take the values MSBFIRST or LSB-FIRST. The dataPin and clockPin must be configured as outputs before using this function. In the example given below, character "A" is shifted out with the least significant bit sent first:

```
dataPin = 12;
clockPin = 13;
value = 'A';
void setup()
{
        pinMode(dataPin, OUTPUT);
        pinMode(clockPin, OUTPUT);
}
shiftOut(dataPin, clockPin, LSBFIRST, value);
```

pulseIn(pin, value, timeout): This function is used to read a pulse present on a pin. If value is HIGH, then the function waits for the pin to become HIGH and starts timing when it becomes HIGH. The timing stops when the pin goes LOW. The length of the time is returned by the function as an integer number. The timeout (in microseconds) is optional, and it specifies for how long the function should wait. An example is given below that waits for pin 3 to go HIGH and the duration is stored in a variable called **tim**:

```
tim = pulseIn(3, HIGH);
```

5.6.5 Time Functions

The following built-in time functions are available:

micros(): This function returns (in microseconds) the number of microseconds since a program started running.
millis(): This function returns (in milliseconds) the number of milliseconds since a program started running.
delay(ms): This function pauses the program for the amount of milliseconds specified in the argument. In the following example, the program is paused for 1 s:
```
delay(1000);
```
delayMicroseconds(μs): This function pauses the program for the amount of microseconds specified in the argument.

5.6.6 Mathematical Functions

The following mathematical functions are available:

min(a, b): Returns the minimum of two numbers. An example is given below that returns 5:

```
Res = min(5, 10);
```

max(a, b): Returns the maximum of two numbers. An example is given below that returns 10:

> Res = max(5, 10);

abs(x): Returns the absolute value of a number. An example is given below that returns number 10:

> A = abs(-10);

constrain(x, min, max): This function returns x if x is between min and max, returns min if x is less than min, and returns max if x is greater than max.

pow(base, exp): This function calculates the value when a number is raised to a power. In the following example, number 2 is raised to power 3 and the returned value is 8:

> r = pow(2, 3);

sqrt(n): Calculates and returns the square root of a number. An example is given below that returns number 4:

> r = sqrt(16);

sq(n): Calculates the square of a number. An example is given below that returns number 16:

> r = sq(4);

5.6.7 Trigonometric Functions

The following trigonometric functions are supported:

sin(n): Calculates the sine of an angle. The angle must be in radians. An example is given below that calculates and returns the sine of 30°:

> a = 30.0*3.14159 / 180.0;
> r = sin(a);

cos(n): Calculates the cosine of an angle. The angle must be in radians. An example is given below that calculates and returns the cosine of 60°:

> a = 60.0*3.14159 / 180.0;
> r = cos(a);

tan(n): Calculates and returns the tangent of an angle. The angle must be in radians. An example is given below that calculates and returns the tangent of 45°:

> a = 45.0*3.14150 / 180.0;
> r = tan(a);

5.6.8 Bit and Byte Functions

The bit and byte functions are useful when we want to extract bits and bytes from variables. The following functions are provided:

lowByte(n): Returns the low byte of a variable.

highByte(n): Returns the high byte of a variable.

bitRead(x, n): Reads a bit of a number starting from the least significant bit, which is bit position 0. Argument n specifies which bit to read. In the following example, bit 3 of number 15 (binary pattern "00001111") is read and is returned as 1:

> r = bitRead(15, 3);

bitWrite(x, n, b): Sets the specified bit of a variable. Argument n specifies the bit number to write to starting from the least significant bit, which is bit 0, and argument b specifies the value (0 or 1) to write to. In the following example, bit 3 of number 7 (binary pattern "00000111") is changed to 1, making the number to be 15 (binary pattern "00001111"):

> bitWrite(7, 3, 1);

bitSet(x, n): Sets bit n of variable x to 1. The bit numbering starts from the least significant bit position that is taken as 0. In the following example, bit 2 of variable tmp is set to 1:

> bitSet(tmp, 2);

bitClear(x, n): Sets bit n of variable x to 0. The bit numbering starts from the least significant bit position that is taken as 0. In the following example, bit 2 of variable tmp is cleared:

> bitClear(tmp, 2);

bit(n): This function returns the value (0 or 1) of a bit in a variable. The least significant bit (bit 0) is 1, bit 1 is 2, bit 2 is 4, and so on.

5.6.9 Interrupt Enable/Disable Functions

Function **interrupts()** is used to enable interrupts. Similarly, function **noInterrupts()** is used to disable interrupts.

5.7 Summary

This chapter presented an introduction to the C language. The basic structure of a C program is given with an explanation of the basic elements that make a C program.

The data types of the C language have been described with examples. Various data structures, such as arrays, structures, and unions, have been described with examples.

Functions are important building blocks of all complex programs. The creation and use of functions in programs have been described with many examples.

Various flow control statements such as *if*, *switch*, *while*, *do*, *break*, and so on, have been explained in the chapter with examples on their use in C programs.

Pointers are important elements of all professional C programs. The basic principles of pointers have been described with various examples to show how they can be used in programs and in functions.

Finally, the basic features of the C language, developed specifically for the PIC32 microcontroller series, have been explained briefly, including a brief description of the built-in functions.

5.8 Exercises

1. What does program repetition mean? Describe how program repetition can be created using **for** and **while** statements.
2. What is an array? Write statements to define arrays with the following properties:
   (a) Array of 10 integers
   (b) Array of 5 floating point numbers
   (c) Array of 3 characters
3. Given that x = 10 and y = 0, list the outcome of the following expressions as either TRUE or FALSE:
   (a) x > 0 && y > 0
   (b) y <= 0
   (c) x == 10 || y == 5
4. Assuming that a = 0x2E and b = 0x6F, determine the results of the following bitwise operations:
   (a) a | b
   (b) b & 0xFF
   (c) a ^ b
   (d) b ^ 0xFF
5. How many times does each of the following loops iterate?
   (a) for(j = 0; j < 5; j++)
   (b) for(i = 0; i <= 5; i++)
   (c) for(i = 10; i > 0; i−−)
   (d) for(i = 10; i > 0; i −= 2)
6. Write a program to calculate the sum of all positive integer numbers between 1 and 5.
7. Write a program to calculate the average of all the numbers from 1 to 100.
8. Derive equivalent **if–else** statements for the following:
   (a) (x > y) ? 1: 0
   (b) (x > y) ? (a − b) : (a + b)

9. What can you say about the following **while** loop?

```
i = 0;
Count = 0;
while (i < 10)
{
        Count++;
}
```

10. What will be the value of **Sum** at the end of the following **do while** loop?

```
k = 0;
Sum = 0;
do
{
        Sum++;
        k++;
}while(k < 20);
```

11. What can you say about the following program?

```
Sum = 0;
for(;;)
{
        Sum++;
}
```

12. Rewrite the following code using a **while** statement:

```
k = 1;
for(i = 1; i <= 10; i++)
{
        k = k + i;
}
```

13. Write a function to receive two integer parameters and to return the larger one to the calling program.
14. Write a program to store the even numbers between 0 and 20 in an integer array. Then call a function to calculate and store the square of each array element in the same array.
15. Write a function to convert inches to centimetres. The function should receive inches and return the equivalent centimetres. Show how you can call this function in a main program to convert 5.2 into centimetres.

16. Write a program to add two matrices **A** and **B**, having dimensions 2×2, and store the result in a third matrix called **C**.

17. Write a function to receive three matrices **A**, **B**, and **C**. Perform the matrix operation $\mathbf{C} = \mathbf{A} \times \mathbf{B}$. Show how the function can be called from a main program.

18. Explain what a function prototype is. Where can it be used?

19. Write a function to convert a two-digit hexadecimal number 0x00 to 0xFF into decimal. Show how you can call this function to convert 0x1E into decimal.

20. In an experiment, the relationship between the x and y values is calculated to be as in the following table:

x	y
1.2	4.0
3.0	5.2
2.4	7.0
4.5	9.1

Write a function using the **switch** statement to find y, given x. Show how the function can be called from a main program to return the value of y when x is 2.4.

21. Explain how you can pass variable number of arguments to a function. Write a function to find the average of the numbers supplied as parameters. Assume that the first parameter is the number of parameters to be supplied. Show how this function can be called from a main program to calculate the average of numbers 2, 3, 5, and 7.

22. Repeat Exercise 19 using **if–else** statements.

CHAPTER 6

Microcontroller Program Development

Chapter Outline

Before writing a program, it is always helpful first to derive the program's algorithm. Although simple programs can easily be developed by writing the code without any prior preparation, the development of complex programs almost always becomes easier if the algorithm is first derived. Once the algorithm is ready, writing of the actual program code is not a difficult task.

A program's algorithm can be described in a variety of graphic and text-based methods, such as flow chart, structure chart, data flow diagram, program description language (PDL), UML activity diagrams, and so on. The problem with graphical techniques is that it can be very time-consuming to draw shapes with text inside them. Also, it is a tedious task to modify an algorithm described using graphical techniques.

Flow charts can be very useful to describe the flow of control and data in small programs where there are only a handful of diagrams, usually not extending beyond a page or two. Flow charts are the earliest software development tools, introduced in 1920s. There are now commercially available flow chart design programs (e.g., SmartDraw, RFFlow, Edraw) that help the user to create and manage large and complex flow charts. One of the problems with flow charts is that the code based on a flow chart tends to be rather unstructured with lots of branches all over the place, and it is difficult to maintain such code. Flow charts are not used nowadays during the development of large programs. Interested readers can find many example flow charts on the Internet.

PIC32 Microcontrollers and the Digilent chipKIT. 978-0-08-099934-0
http://dx.doi.org/10.1016/B978-0-08-099934-0.00006-5

The PDL (sometimes called program design language) can be useful to describe the flow of control and data in small- to medium-size programs. The main advantage of the PDL description is that it is very easy to modify a given PDL since it consists of only text. In addition, the code generated from a PDL-based program is normally structured and easy to maintain.

In this book, we will mainly be using the PDL, but the flow charts will also be given where it is felt to be useful. The next sections briefly describe the basic building blocks of the PDL and flow charts. It is left to the readers to decide which methods to use during the development of their programs.

6.1 Using the Program Description Language and Flow Charts

PDL is free-format English-like text that describes the flow of control and data in a program. It is important to realise that PDL is not a programming language, but a collection of some keywords that enable a programmer to describe the operation of a program in a stepwise and logical manner. In this section, we will look at the basic PDL statements and their flow chart equivalents. The superiority of the PDL over flow charts will become obvious when we have to develop medium-size to large programs.

6.1.1 BEGIN–END

Every PDL program description should start with a BEGIN statement and end with an END statement. The keywords in a PDL description should be highlighted (e.g., bold) to make the reading easier. The program statements should be indented and described between the PDL keywords. An example is shown in Figure 6.1 together with the equivalent flow diagram. Notice that the flow chart also uses BEGIN and END keywords.

6.1.2 Sequencing

For normal sequencing, the program statements should be written in English text and describe the operations performed one after the other. An example is shown in Figure 6.2 together with the equivalent flow chart. It is clear from this example how much easier it is to describe the sequence using PDL.

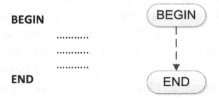

Figure 6.1: BEGIN–END Statement and Equivalent Flow Chart

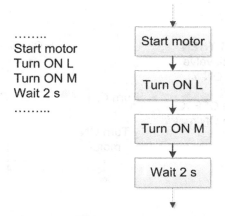

....
Start motor
Turn ON L
Turn ON M
Wait 2 s
....

Figure 6.2: Sequencing and Equivalent Flow Chart

6.1.3 IF–THEN–ELSE–ENDIF

The IF, THEN, ELSE, and ENDIF should be used to conditionally change the flow of control in a program. Every IF line should be terminated with THEN, and every IF block should be terminated with ENDIF statements. Use of the ELSE statement is optional and depends on the application. Figure 6.3 shows an example of using IF–THEN–ENDIF statements, while Figure 6.4 shows the use of IF–THEN–ELSE–ENDIF statements in a program and their equivalent flow charts. Again, the simplicity of using PDL is apparent from this example.

6.1.4 DO–ENDDO

The DO–ENDDO statements should be used when it is required to create iterations, or conditional or unconditional loops in programs. Every DO statement should be terminated

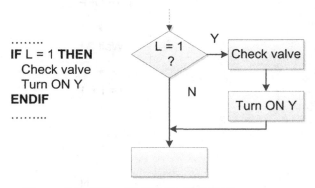

....
IF L = 1 **THEN**
  Check valve
  Turn ON Y
ENDIF
....

Figure 6.3: Using IF–THEN–ENDIF Statements

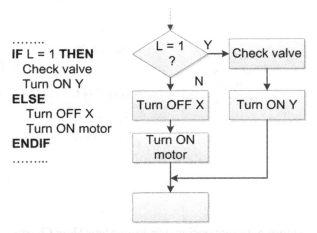

```
........
IF L = 1 THEN
    Check valve
    Turn ON Y
ELSE
    Turn OFF X
    Turn ON motor
ENDIF
........
```

Figure 6.4: Using IF–THEN–ELSE–ENDIF Statements

with an ENDDO. Other keywords, such as FOREVER or WHILE, can be used after the DO statement to indicate an endless loop or a conditional loop, respectively. Figure 6.5 shows an example of a DO–ENDDO loop executed 10 times. Figure 6.6 shows an endless loop created using the FOREVER statement. The flow chart equivalents are also shown in the figures.

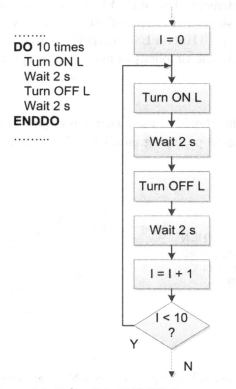

```
........
DO 10 times
    Turn ON L
    Wait 2 s
    Turn OFF L
    Wait 2 s
ENDDO
........
```

Figure 6.5: Using DO–ENDDO Statements

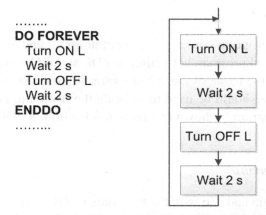

....
DO FOREVER
   Turn ON L
   Wait 2 s
   Turn OFF L
   Wait 2 s
ENDDO
....

Figure 6.6: Using DO–FOREVER Statements

6.1.5 REPEAT–UNTIL

REPEAT–UNTIL is similar to DO–WHILE; the difference is that the statements enclosed by the REPEAT–UNTIL block are executed at least once, while the statements enclosed by the DO–WHILE block may not execute at all if the condition is not satisfied just before entering the DO statement. An example is shown in Figure 6.7, with the equivalent flow chart.

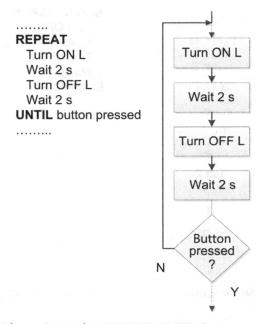

....
REPEAT
   Turn ON L
   Wait 2 s
   Turn OFF L
   Wait 2 s
UNTIL button pressed
....

Figure 6.7: Using REPEAT–UNTIL Statements

6.1.6 Calling Subprograms

In some applications, a program consists of a main program and a number of subprograms (or functions). A subprogram activation in PDL should be shown by adding the CALL statement before the name of the subprogram. In flow charts, a rectangle with vertical lines at each side should be used to indicate the invocation of a subprogram. An example call to a subprogram is shown in Figure 6.8 for both a PDL description and a flow chart.

6.1.7 Subprogram Structure

A subprogram should begin and end with the keywords BEGIN/name and END/name, respectively, where *name* is the name of the subprogram. In flow chart representation, a horizontal line should be drawn inside the BEGIN box and the name of the subprogram should be written in the lower half of the box. An example subprogram structure is shown in Figure 6.9 for both a PDL description and a flow chart.

6.2 Examples

Some examples are given in this section to show how the PDL and flow charts can be used in program development.

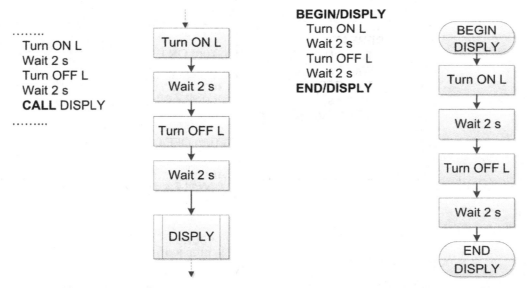

Figure 6.8: Calling a Subprogram   Figure 6.9: Subprogram Structure

Example 6.1

It is required to a write a program to convert hexadecimal numbers "A" to "F" into decimal. Show the algorithm using a PDL and also draw the flow chart. Assume that the number to be converted is called HEX_NUM and the output number is called DEC_NUM.

Solution 6.1

The required PDL is:

```
BEGIN
      IF HEX_NUM = "A" THEN
            DEC_NUM = 10
      ELSE IF HEX_NUM = "B" THEN
            DEC_NUM = 11
      ELSE IF HEX_NUM = "C" THEN
            DEC_NUM = 12
      ELSE IF HEX_NUM = "D" THEN
            DEC_NUM = 13
      ELSE IF HEX_NUM = "E" THEN
            DEC_NUM = 14
      ELSE IF HEX_NUM = "F" THEN
            DEC_NUM = 15
      ENDIF
END
```

The required flow chart is shown in Figure 6.10. Notice that it is much easier to write the PDL statements than drawing the flow chart shapes and writing text inside them.

Example 6.2

The PDL of part of a program is given as follows:

```
J = 0
M = 0
DO WHILE J < 10
      DO WHILE M < 20
            Flash the LED
            Increment M
      ENDDO
      Increment J
ENDDO
```

Show how this PDL can be implemented by drawing a flow chart.

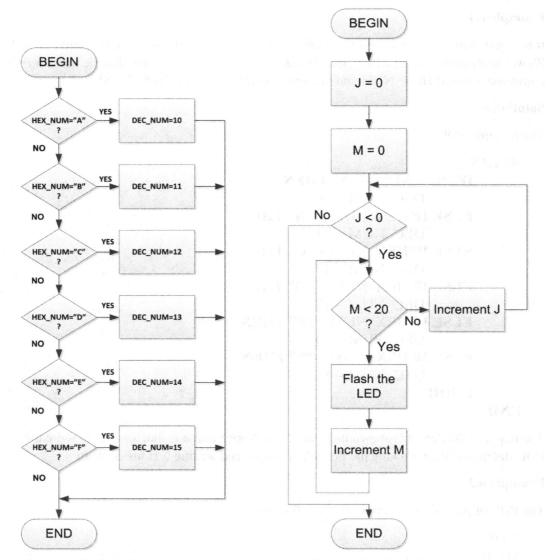

Figure 6.10: Flow Chart Solution of
Example 6.1

Figure 6.11: Flow Chart Solution of
Example 6.2

Solution 6.2

The required flow chart is shown in Figure 6.11. Here again notice how complicated the flow chart can be even for a simple nested DO–WHILE loop.

Example 6.3

It is required to write a program to calculate the sum of integer numbers between 1 and 100. Show the algorithm using a PDL and also draw the flow chart. Assume that the sum will be stored in a variable called SUM.

Solution 6.3

The required PDL is:

```
BEGIN
        SUM = 0
        I = 1
        DO 100 TIMES
                SUM = SUM + I
                Increment I
        ENDDO
END
```

The required flow chart is shown in Figure 6.12.

Example 6.4

It is required to write a program to calculate the sum of all the even numbers between 1 and 10 inclusive. Show the algorithm using a PDL and also draw the flow chart. Assume that the sum will be stored in a variable called SUM.

Solution 6.4

The required PDL is:

```
BEGIN
        SUM = 0;
        CNT = 1
        REPEAT
                IF CNT is even number THEN
                        SUM = SUM + CNT
                ENDIF
                INCREMENT CNT
        UNTIL CNT > 10
END
```

The required flow chart is shown in Figure 6.13. Notice how complicated the flow chart can be for a very simple problem such as this.

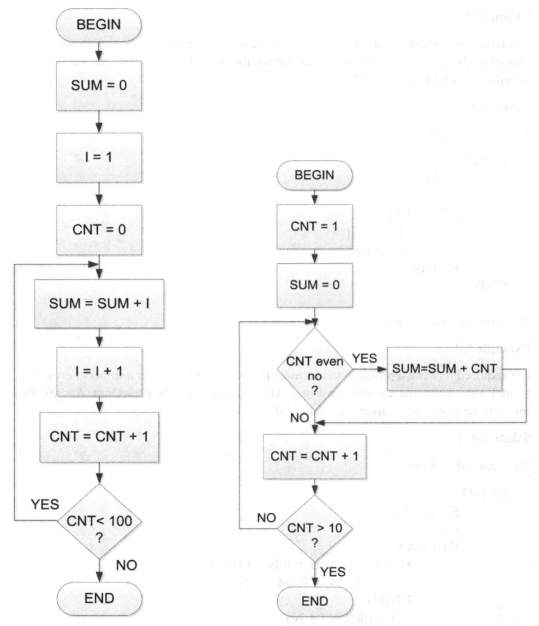

Figure 6.12: Flow Chart Solution of Example 6.3

Figure 6.13: Flow Chart Solution of Example 6.4

6.3 Representing for Loops in Flow Charts

Most programs include some form of iteration or looping. One of the easiest ways to create a loop in a C program is by using the *for* statement. This section shows how a *for* loop can be represented in a flow chart. As shown below, there are several methods of representing a *for* loop in a flow chart.

Suppose that we have a *for* loop as below and we wish to draw an equivalent flow chart:

```
for(m = 0; m < 10; m++)
{
        Cnt = Cnt + 2*m;
}
```

6.3.1 Method 1

Figure 6.14 shows one of the methods for representing the above *for* loop as with a flow chart. Here, the flow chart is drawn using the basic primitive components.

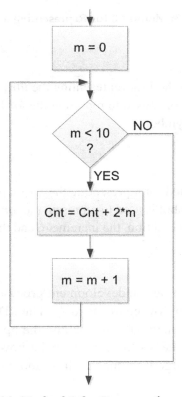

Figure 6.14: Method 1 for Representing a for Loop

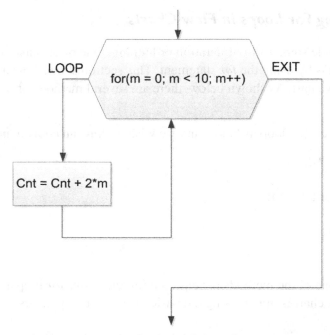

Figure 6.15: Method 2 for Representing a *for* Loop

6.3.2 Method 2

Figure 6.15 shows the second method for representing the *for* loop with a flow chart. Here, a hexagon-shaped flow chart symbol is used to represent the *for* loop and the complete *for* loop statement is written inside this symbol.

6.3.3 Method 3

Figure 6.16 shows the third method for representing the *for* loop with a flow chart. Here again a hexagon-shaped flow chart symbol is used to represent the *for* loop and the symbol is divided into three to represent the initial condition, the increment, and the terminating condition.

6.4 Summary

This chapter has described the program development process using the Program description language (PDL) and flow charts as tools. The PDL is commonly used as it is a simple and convenient method of describing the operation of a program. It consists of several English-like keywords. Although the flow chart is also a useful tool, it can be very tedious in large programs to draw and modify shapes and write text inside them.

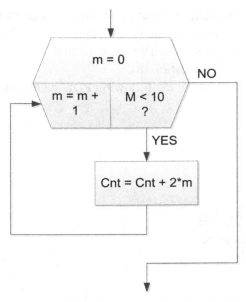

Figure 6.16: Method 3 for Representing a *for* Loop

6.5 Exercises

1. Describe the various shapes used in drawing flow charts.
2. Describe how the various keywords used in PDL can be used to describe the operation of a program.
3. What are the advantages and disadvantages of flow charts?
4. It is required to write a program to calculate the sum of numbers from 1 to 10. Draw a flow chart to show the algorithm for this program.
5. Write the PDL statements for question (4).
6. It is required to write a program to calculate the roots of a quadratic equation, given the coefficients. Draw a flow chart to show the algorithm for this program.
7. Write the PDL statements for question (6).
8. Draw the equivalent flow chart for the following PDL statements:

   DO WHILE count < 10
         Increment J
         Increment count
   ENDDO

9. It is required to write a function to calculate the sum of numbers from 1 to 10. Draw a flow chart to show how the function subprogram and the main program can be implemented.
10. Write the PDL statements for question (9).

11. It is required to write a function to calculate the cube of a given integer number and then call this function from a main program. Draw a flow chart to show how the function subprogram and the main program can be implemented.

12. Write the PDL statements for question (8).

13. Draw the equivalent flow chart for the following PDL statements:

```
J = 0
K = 0
REPEAT
        Flash LED A
        Increment J
        REPEAT
                Flash LED B
                Increment K
        UNTIL K = 10
UNTIL J > 15
```

CHAPTER 7

Simple chipKIT MX3—Based Projects

Chapter Outline

PIC32 Microcontrollers and the Digilent chipKIT. 978-0-08-099934-0
http://dx.doi.org/10.1016/B978-0-08-099934-0.00007-7

In this chapter, we shall be looking at the design of simple PIC32 32-bit microcontroller-based projects using the chipKIT MX3 development board and the MPIDE development environment, with the idea of becoming familiar with basic interfacing techniques and learning how to use the various microcontroller peripheral registers. We will look at the design of projects using LEDs, push-button switches, keyboards, LED arrays, sound devices, and so on. It is recommended that the reader move through the projects in their given order. The following are provided for each project:

- Description of the project
- Block diagram of the project
- Circuit diagram of the project
- Description of the hardware
- Algorithm description [in program description language (PDL)]
- Program listing
- Photos of the project (where applicable)
- Suggestions for further development (where applicable)

In this book, we will be using the PDL for all the projects.

The use of the MPIDE development environment will be discussed in detail in Project 7.1.

7.1 Project 7.1 – Flashing LED

7.1.1 Project Description

This is a simple project that flashes one of the LEDs on the chipKIT MX3 development board at a rate of once a second. There are two LEDs on the development board, and in this project the LED connected to I/O port pin RF0 (LED named LD4) is used. This LED has the logical pin number 42 (see Table 5.8).

An LED can be connected to a microcontroller output port in two different modes: *current-sinking* mode and *current-sourcing* mode.

7.1.2 Current-Sinking

As shown in Figure 7.1, in current-sinking mode the anode leg of the LED is connected to the Vdd supply, and the cathode leg is connected to the microcontroller output port through a current-limiting resistor.

The voltage drop across an LED is around 2 V. The brightness of the LED depends on the current through the LED, and this current can vary between a few milliamperes to 16 mA depending on the type of LED used.

The LED is turned ON when the output of the microcontroller is at logic 0 so that current flows through the LED. We can calculate the value of the required resistor as follows:

$$R = \frac{\text{Vdd} - V_{\text{LED}}}{I_{\text{LED}}}$$

where Vdd is the supply voltage (3.3 V); V_{LED} is the voltage drop across the LED (2 V); I_{LED} is the current through the LED (say, 10 mA).

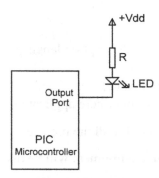

Figure 7.1: LED Connected in Current-Sinking Mode

Substituting the values into the equation, we get

$$R = \frac{3.3 - 2}{10} = 130 \ \Omega$$

The nearest physical resistor is 120 Ω.

7.1.3 Current-Sourcing

As shown in Figure 7.2, in current-sourcing mode the anode leg of the LED is connected to the microcontroller output port and the cathode leg is connected to the ground through a current-limiting resistor.

In this mode, the LED is turned ON when the microcontroller output port is at logic 1, that is, Vdd. In practice, the output voltage is about 3.3 V and the value of the resistor can be determined as:

$$R = \frac{\text{Vdd} - V_{\text{LED}}}{I_{\text{LED}}}$$

which gives the same resistor value of 120 Ω.

7.1.4 Project Hardware

In this project, the LED named LD4 on the chipKIT MX3 development board is used. This LED is connected to port pin RF0 through a switching transistor as shown in Figure 4.4 (it is also possible to connect the LED through a current-limiting resistor as shown in Figures 7.1 and 7.2).

7.1.5 Project PDL

The operation of the project is described in the PDL given in Figure 7.3. At the beginning of the program, the LED port (logical port 42) is configured as output. The program then

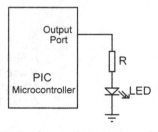

Figure 7.2: LED Connected in Current-Sourcing Mode

BEGIN
    Configure logical I/O pin 42 (port pin RF0) as output
    DO FOREVER
        Send 1 to the port pin (turn LED ON)
        Wait 1 second
        Send 0 to the port pin (turn LED OFF)
        Wait 1 second
    ENDDO
END

Figure 7.3: PDL of Project

flashes the LED at a rate of 1 s by sending logic 1 and then logic 0 to the LED with 1-s delay between each output.

7.1.6 Project Program

The program is called LED1, and the program listing is shown in Figure 7.4. At the beginning of the program, the port pin where LD4 is connected to is configured as an output inside the

```
/*==========================================================================
                          LED FLASHING PROGRAM
                          ====================

This program flashes the LED connected to I/O pin 42 (RF0) of the chipKIT MX3 development
Board every second.


Author:        Dogan Ibrahim
Date:          May, 2014
File:          LED1
Board:         chipKIT MX3
===========================================================================*/
int ledPin = 42;                              // LED port

void setup()                                  // initialization
{
pinMode(ledPin, OUTPUT);                      // set digital pin as output
}


void loop()
{
        digitalWrite(ledPin, HIGH);           // turn LED ON
        delay(1000);                          // wait 1 second
        digitalWrite(ledPin, LOW);            // turn LED OFF
        delay(1000);                          // wait 1 second
}
```

Figure 7.4: Program Listing

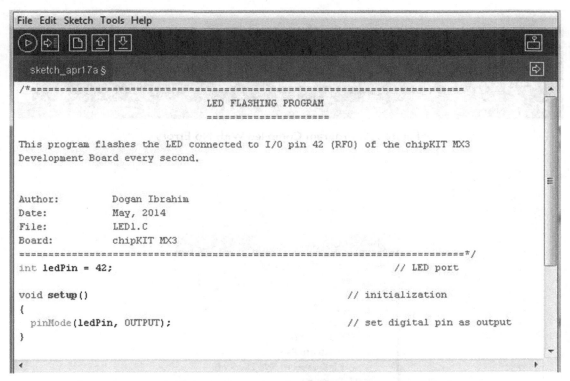

Figure 7.5: Enter the Program Using the MPIDE Development Environment

setup() function. The LED is then flashed ON and OFF by sending 1 and 0 continuously with 1-s delay between each output.

The following steps describe how to create the program and upload it to the flash memory of the target microcontroller on the chipKIT MX3 development board:

Step 1: Start the MPIDE development environment.

Step 2: Enter the program shown in Figure 7.4 (see Figure 7.5).

Step 3: Select the development board type by clicking *Tools → Board → Cerebot → Mx3cK*.

Step 4: Compile the program by clicking the *Verify* button in the toolbar. If the program compiles successfully, you should get no error messages (see Figure 7.6).

Step 5: Connect the chipKIT MX3 development board to the USB port of your computer. You should see the red power LED coming on and the green BootLoader LED (LD4) flashing rapidly.

Step 6: Select the serial COM port number assigned to the board. This number can be found from *Control Panel → System → Device Manager* under the heading *Ports (COM & LPT) → USB Serial Port* (see Figure 7.7).

Click *Tools → Serial Port* to select or confirm the assigned port as shown in Figure 7.8.

Done compiling.

Binary sketch size: 6008 bytes (of a 126976 byte maximum)

6   chipKIT UNO32 on COM1

Figure 7.6: Program Compiled With No Errors

Device Manager

File   Action   View   Help

▲ Dogan-HP
  ▷ Batteries
  ▷ Biometric Devices
  ▲ Bluetooth Radios
      Broadcom 2070 Bluetooth
  ▷ Computer
  ▷ Dataram RAMDisk Devices
  ▷ Disk drives
  ▷ Display adapters
  ▷ DVD/CD-ROM drives
  ▷ Human Interface Devices
  ▷ IDE ATA/ATAPI controllers
  ▷ Imaging devices
  ▷ Keyboards
  ▷ Memory technology driver
  ▷ Mice and other pointing devices
  ▷ Monitors
  ▷ Network adapters
  ▲ Ports (COM & LPT)
      USB Serial Port (COM28)
  ▷ Processors
  ▷ Sound, video and game controllers
  ▷ System devices
  ▷ Universal Serial Bus controllers

Figure 7.7: The Assigned COM Port Is COM28

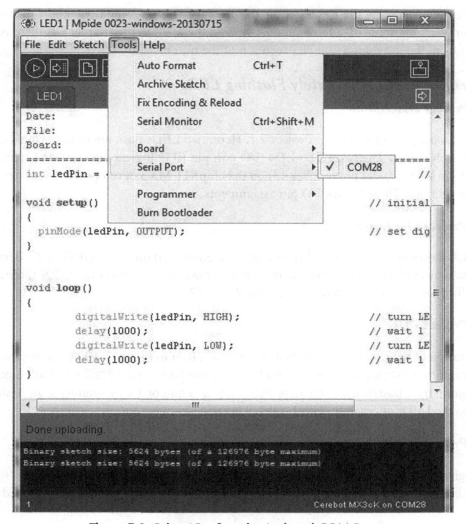

Figure 7.8: Select/Confirm the Assigned COM Port

Step 7: Upload the program to the target microcontroller by clicking *File → Upload to I/O Board*. You should see the message *Done uploading* as in Figure 7.9.
Step 8: The microcontroller should reset automatically and the program execution should start. You should now see LED LD4 flashing continuously at a rate of 1 s.

Done uploading.

Binary sketch size: 5624 bytes (of a 126976 byte maximum)

Figure 7.9: Successful Uploading of the Program

Notice that after pressing the Reset button, the green LED flashes rapidly and the micro-controller waits for about 10 s to communicate with the MPIDE programmer. If there is no response from the programmer, then execution of the user program starts.

7.2 Project 7.2 – Alternately Flashing LEDs

7.2.1 Project Description

This is a simple project similar to Project 7.1. Here, two LEDs flash alternately with 1-s delay between each output. The two LEDs LD4 (I/O port pin RF0, logical pin number 42) and LD5 (I/O port pin RF1, logical pin number 43) on the chipKIT MX3 development board are used in this project. See Table 5.8 for I/O pin assignments.

7.2.2 Project Hardware

Figure 4.4 shows how the LEDs LD4 and LD5 are connected on the chipKIT MX3 develop-ment board through switching transistors (it is also possible to connect the LEDs through current-limiting resistors as shown in Figures 7.1 and 7.2).

7.2.3 Project PDL

The operation of the project is described in the PDL given in Figure 7.10. At the beginning of the program, LD4 (logical port 42) and LD5 (logical port 43) are configured as outputs. The program then flashes the two LEDs alternately at a rate of 1 s by sending logic 1 and then logic 0 to the LEDs alternately with 1-s delay between each output.

7.2.4 Project Program

The program is called LED2, and the program listing is shown in Figure 7.11. At the begin-ning of the program, the port pins where LD4 and LD5 are connected to are configured as

```
BEGIN
        Configure logical I/O pin 42 (port pin RF0) as output
        Configure logical I/O pin 43 (port pin RF1) as output
        Turn OFF LD4 and LD5
        DO FOREVER
                Send 1 to port pin RF0 (turn LD4 ON)
                Send 0 to port pin RF1 (turn LD5 OFF)
                Wait 1 second
                Send 0 to port pin RF0 (turn LD4 OFF)
                Send 1 to port pin RF1 (turn LD5 ON)
                Wait 1 second
        ENDDO
END
```

Figure 7.10: PDL of Project

```
/*=========================================================================
                        LED FLASHING PROGRAM
                        ====================

This program flashes alternately two LED connected to I/O pin 42 (RF0) and I/O pin 43 (RF1)
of the chipKIT MX3 development board every second.


Author:         Dogan Ibrahim
Date:           May, 2014
File:           LED2
Board:          chipKIT MX3
=========================================================================*/
int LD4 = 42;                                   // LED LD4 port
int LD5 = 43;                                   // LED LD5 port

void setup()                                    // initialization
{
  pinMode(LD4, OUTPUT);                         // set LD4 pin as output
  pinMode(LD5, OUTPUT);                         // set LD5 pin as output
  digitalWrite(LD4, LOW);                       // turn OFF LD4
  digitalWrite(LD5, LOW);                       // turn OFF LD5
}


void loop()
{
        digitalWrite(LD4, HIGH);                // turn LD4 ON
        digitalWrite(LD5, LOW);                 // turn LD5 OFF
        delay(1000);                            // wait 1 second
        digitalWrite(LD4, LOW);                 // turn LD4 OFF
        digitalWrite(LD5, HIGH);                // turn LD5 ON
        delay(1000);                            // wait 1 second
}
```

Figure 7.11: Program Listing

outputs inside the *setup()* function. Also, both LEDs are turned OFF to start with. The LEDs are then flashed ON and OFF alternately by sending 1 and 0 continuously with 1-s delay between each output.

7.3 Project 7.3 – Lighthouse Flashing LED

7.3.1 Project Description

This is a simple project using LED LD4. In this project, the LED is flashed in a group of two quick flashes every second. The flashing rate is assumed to be 200 ms. This type of flashing is identified as Gp Fl(2) in maritime lighthouse lights. Thus, the flashing is repeated as follows:

LED ON
Wait 200 ms
LED OFF
Wait 100 ms
LED ON
Wait 200 ms
LED OFF
Wait 100 ms
Wait 400 ms

7.3.2 Project Hardware

Figure 4.4 shows how LD4 is connected on the chipKIT MX3 development board through a switching transistor (it is also possible to connect the LEDs through current-limiting resistors as shown in Figures 7.1 and 7.2).

7.3.3 Project PDL

The operation of the project is described in the PDL given in Figure 7.12. At the beginning of the program, LD4 (logical port 42) is configured as an output. The program then flashes the LED as a lighthouse signal Gp Fl(2) as described above.

7.3.4 Project Program

The program is called LED3, and the program listing is shown in Figure 7.13. At the beginning of the program, the port pin where LD4 is connected to is configured as an output inside

```
BEGIN
        Configure logical I/O pin 42 (port pin RF0) as output
        DO FOREVER
                Send 1 to port pin RF0 (turn LD4 ON)
                Wait 200ms
                Send 0 to port pin RF0 (turn LD4 OFF)
                Wait 100ms
                Send 1 to port pin RF0 (turn LD4 ON)
                Wait 200ms
                Send 0 to port pin RF0 (turn LD4 OFF)
                Wait 100ms
                Wait 400ms
        ENDDO
END
```

Figure 7.12: PDL of Project

```
/*========================================================================
                        LED LIGHTHOUSE FLASHING PROGRAM
                        ================================

This program flashes the LED connected to I/O pin 42 (RF0) as a lighthouse signal GpFl(2)


Author:         Dogan Ibrahim
Date:           May, 2014
File:           LED3
Board:          chipKIT MX3
========================================================================*/
int LD4 = 42;                               // LED LD4 port

void setup()                                // initialization
{
  pinMode(LD4, OUTPUT);                     // set LD4 pin as output
}


void loop()
{
        digitalWrite(LD4, HIGH);            // turn LD4 ON
        delay(200);                         // wait 200ms
        digitalWrite(LD4, LOW);             // turn LD4 OFF
        delay(100);                         // wait 100ms
        digitalWrite(LD4, HIGH);            // turn LD4 ON
        delay(200);                         // wait 200ms
        digitalWrite(LD4, LOW);             // turn LD4 OFF
        delay(100);                         // wait 100ms
        delay(400);                         // wait 400ms
}
```

Figure 7.13: Program Listing

the *setup()* function. The LED is then flashed ON and OFF to simulate the lighthouse signalling Gp Fl(2) as described above.

7.4 Project 7.4 – LED With Push-Button Switch

7.4.1 Project Description

This is a simple project using LED LD4 with a push-button switch. In this project, the LED is controlled from a push-button switch and is turned ON and OFF when the switch is pushed and released, respectively.

7.4.2 Block Diagram

The block diagram of the project is shown in Figure 7.14.

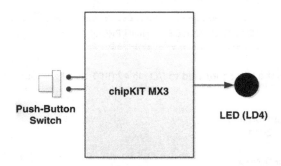

Figure 7.14: Block Diagram of the Project

7.4.3 Project Hardware

In this project, the four-button Pmod module PmodBTN is used (see Figure 7.15). This module consists of four push-button switches labelled BTN0–BTN3 with on-board debounce filters as shown in Figure 7.16.

The PmodBTN module has a six-pin header and is connected to Pmod connector JA on the chipKIT MX3 development board.

The pin connections of the PmodBTN module are as follows:

Pin Number	Pin Function
1	BTN0
2	BTN1
3	BTN2
4	BTN3
5	GND
6	VCC

A switch pin is at logic 0 and goes to logic 1 when the switch is pressed. When connected to Pmod connector JA, the interface between the switch and the

Figure 7.15: PmodBTN Module

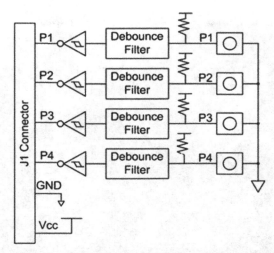

Figure 7.16: PmodBTN Module Connection Diagram

microcontroller I/O pins when the PmodBTN is connected to the top row of the connector is as follows:

Pin Number	Microcontroller I/O Port Pin
1	RE0 (JA-01, logical I/O pin 0)
2	RE1 (JA-02, logical I/O pin 1)
3	RE2 (JA-03, logical I/O pin 2)
4	RE3 (JA-04, logical I/O pin 3)

In this project, BTN0 is used. This switch has the logical I/O pin number 0 as shown in the above table.

Figure 7.17 shows the PmodBTN module connected to the development board.

7.4.4 Project PDL

The operation of the project is described in the PDL given in Figure 7.18. At the beginning of the program, LD4 (logical port 42) is configured as an output and connector JA-01 pin (logical port 0) is configured as an input. The LED is turned ON when the push-button switch is pressed, that is, when the **ButtonState** is at logic 1.

7.4.5 Project Program

The program is called PUSH1, and the program listing is shown in Figure 7.19. At the beginning of the program, the port pin where LED LD4 is connected to is configured as an output and the port pin where button BTN0 is connected to is configured as an input inside the

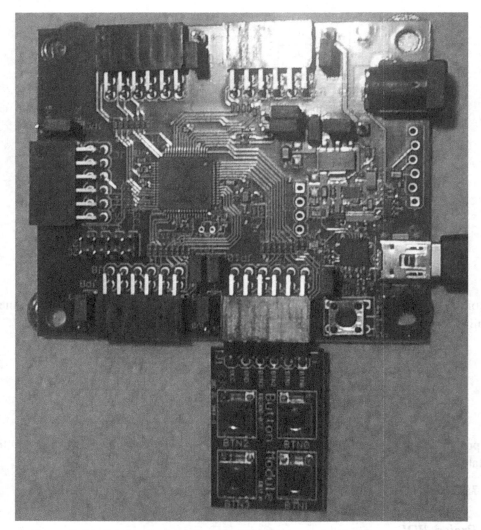

Figure 7.17: Connecting the PmodBTN Module to the Development Board

setup() function. The program then reads the state of the button, and the LED is turned ON whenever the button is pressed; otherwise, the LED is turned OFF.

7.5 Project 7.5 – Wait Before Flashing the LED

7.5.1 Project Description

This is a simple project similar to Project 7.4. Here, LED LD4 and push-button BTN0 are used as in Project 7.4. THE LED is normally OFF and starts flashing at a rate of 1 s when the push-button switch is pressed.

BEGIN
    Configure logical I/O pin 42 (port pin RF0, LD4) as output
    Configure logical I/O pin 0 (port pin RE0, connector JA-01) as input
    DO FOREVER
        Read Button state
        IF push-button switch is pressed **THEN**
            Turn ON LED LD4
        ELSE
            Turn OFF LED LD4
        ENDIF
    ENDDO
END

Figure 7.18: PDL of Project

```
/*==========================================================================
                        LED WITH PUSH-BUTTON SWITCH
                        ===========================

This program uses a push-button switch (connected to logical port 0) and
an LED (connected to logical port 42). The LED is controlled by the push-button
switch. The LED turns ON and OFF when the switch is pushed and released respectively


Author:        Dogan Ibrahim
Date:          May, 2014
File:          PUSH1
Board:         chipKIT MX3
===========================================================================*/
int LD4 = 42;                              // LD4 logical port number
int BTN0 = 0;                              // BTN0 logical port number
int ButtonState;

void setup()
{
 pinMode(LD4, OUTPUT);                     // set LD4 as output
 pinMode(BTN0, INPUT);                     // set BTN0 as input
 digitalWrite(LD4, LOW);                   // turn LD4 OFF to start with
}


void loop()
{
 ButtonState = digitalRead(BTN0);         // Read state of the button

 if(ButtonState == 1)                     // if BTN0 is pressed
    digitalWrite(LD4, HIGH);              // turn LD4 ON
 else
    digitalWrite(LD4, LOW);               // turn LD4 OFF
}
```

Figure 7.19: Program Listing of Project

```
        BEGIN
               Configure logical I/O pin 42 (port pin RF0, LD4) as output
               Configure logical I/O pin 0 (port pin RE0, connector JA-01) as input
               Turn OFF the LED
               DO UNTIL the button is pressed
                       Wait
               ENDDO
               DO FOREVER
                       Turn ON the LED
                       Wait 1 second
                       Turn OFF the LED
                       Wait 1 second
               ENDDO
        END
```

Figure 7.20: PDL of Project

7.5.2 Block Diagram

The block diagram of the project is as in Figure 7.14.

7.5.3 Project Hardware

The project hardware is as in Project 7.4 where a PmodBTN module is used. This module consists of four push-button switches labelled BTN0–BTN3 with on-board debounce filters, and BTN0 is used in this project. The PmodBTN module is connected to Pmod connector JA, and thus BTN0 has the logical I/O pin number 0 as in the previous project.

7.5.4 Project PDL

The operation of the project is described in the PDL given in Figure 7.20. At the beginning of the program, LD4 (logical port 42) is configured as an output and connector JA-01 pin (logical port 0) is configured as an input. The LED is normally OFF and starts flashing at a rate of 1 s as soon as the switch is pressed.

7.5.5 Project Program

The program is called PUSH2, and the program listing is shown in Figure 7.21. At the beginning of the program, the port pin where LED LD4 is connected to is configured as an output and the port pin where button BTN0 is connected to is configured as an input inside the *setup()* function. The program then reads the state of the button in a loop and waits until the button is pressed. As soon as the button is pressed, the LED starts flashing at a rate of 1 s.

```
/*========================================================================
                        LED WITH PUSH-BUTTON SWITCH
                        ===========================

This program uses a push-button switch (connected to logical port 0) and
an LED (connected to logical port 42). The LED is controlled by the
push-button switch. The LED turns ON and OFF when the switch is pushed and
released respectively.


Author:         Dogan Ibrahim
Date:           May, 2014
File:           PUSH2
Board:          chipKIT MX3
========================================================================*/
int LD4 = 42;                                    // LD4 logical port number
int BTN0 = 0;                                    // BTN0 logical port number
int ButtonState = 0;

void setup()
{
        pinMode(LD4, OUTPUT);                    // set LD4 as output
        pinMode(BTN0, INPUT);                    // set BTN0 as input
        digitalWrite(LD4, LOW);                  // turn LD4 OFF to start with
}


void loop()
{
while(ButtonState == 0)                          // Waits until button is pressed
{
        ButtonState = digitalRead(BTN0);
}

while(1)
{
        digitalWrite(LD4, HIGH);                 // turn LD4 ON
        delay(1000);                             // wait 1 second
        digitalWrite(LD4, LOW);                  // turn LD4 OFF
        delay(1000);                             // wait 1 second
}
}
```

Figure 7.21: Program Listing of the Project

7.6 Project 7.6 – LED With Two Push-Button Switches

7.6.1 Project Description

This is a simple project similar to Project 7.4. Here, LED LD4 and two push-button switches BTN0 and BTN1 are used. The LED is normally OFF and starts flashing at a rate of 500 ms

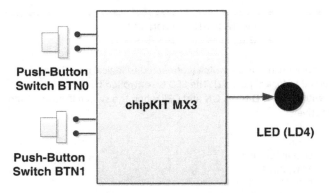

Figure 7.22: Block Diagram of the Project

when the push-button switch BTN0 is pressed. The LED stops flashing when push-button BTN1 is pressed.

7.6.2 Block Diagram

The block diagram of the project is shown in Figure 7.22.

7.6.3 Project Hardware

The project hardware is similar to Project 7.4 where a PmodBTN module is used. This module consists of four push-button switches labelled BTN0–BTN3 with on-board debounce filters. BTN0 is used to start flashing the LED, and BTN1 is used to stop flashing the LED. The PmodBTN module is connected to Pmod connector JA. BTN0 (connector pin JA-01) and BTN1 (connector pin JA-02) have the logical I/O pin numbers 0 and 1, respectively (see Project 7.4).

7.6.4 Project PDL

The operation of the project is described in the PDL shown in Figure 7.23. At the beginning of the program, LD4 (logical port 42) is configured as an output and connector JA-01 pin (BTN0, logical port 0) and JA-02 pin (BTN1, logical port 1) are configured as inputs. The LED is turned ON when the push-button switch BTN0 is pressed. The LED is turned OFF when push-button switch BTN1 is pressed.

7.6.5 Project Program

The program is called PUSH2, and the program listing is shown in Figure 7.24. At the beginning of the program, the port pin where LED LD4 is connected to is configured as an output and the port pins where buttons BTN0 and BTN1 are connected to are configured as inputs

```
BEGIN
        Configure logical I/O pin 42 (port pin RF0, LD4) as output
        Configure logical I/O pin 0 (BTN0, connector JA-01) as input
        Configure logical I/O pin 1 (BTN1, connector JA-02) as input
        DO FOREVER
                IF BTN0 is pressed THEN
                        Flash = 1
                END
                IF BTN1 is pressed THEN
                        Flash = 0
                END
                IF Flash = 1 THEN
                        Turn ON LED LD4
                        Wait 0.5 second
                        Turn OFF LED LD4
                        Wait 0.5 second
                ENDIF
        ENDDO
END
```

Figure 7.23: PDL of the Project

inside the *setup()* function. The program then reads the state of both buttons. If BTN0 is pressed, then variable **Flash** is set to 1. If, on the other hand, BTN1 is pressed, then variable **Flash** is cleared to zero. The LED starts flashing at a rate of 500 ms if variable **Flash** is set to 1. The flashing stops when variable **Flash** is cleared.

7.7 Project 7.7 – Rotating LEDs

7.7.1 Project Description

In this project, four LEDs are used and these LEDs are turned ON in a rotating manner, that is, the ON/OFF pattern of the LEDs is as follows:

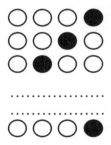

7.7.2 Block Diagram

The block diagram of the project is shown in Figure 7.25.

```
/*=========================================================================
                  LED WITH TWO PUSH-BUTTON SWITCHES
                  ==================================
```

This program uses two push-button switches BTN0 and BTN1 with PmodBTN module.
LED LD4 starts flashing when BTN0 is pressed. The flashing stops when BTN1 is pressed.
The logical port numbers of ports used are:

LD4   – port 42
BTN0 – port 0
BTN1 – port 1

```
Author:          Dogan Ibrahim
Date:            May, 2014
File:            PUSH2
Board:           chipKIT MX3
=========================================================================*/
int LD4 = 42;                               // LD4 logical port number
int BTN0 = 0;                               // BTN0 logical port number
int BTN1 = 1;                               // BTN1 logical port number
int BTN0State, BTN1State;
int Flash = 0;

void setup()
{
        pinMode(LD4, OUTPUT);               // set LD4 as output
        pinMode(BTN0, INPUT);               // set BTN0 as input
        pinMode(BTN1, INPUT);               // set BTN1 as input
        digitalWrite(LD4, LOW);             // turn LD4 OFF to start with
}


void loop()
{
        BTN0State = digitalRead(BTN0);      // read state of BTN0
        BTN1State = digitalRead(BTN1);      // read state of BTN1
        if(BTN0State == 1)Flash = 1;        // if BTN0 is pressed
        if(BTN1State == 1) Flash = 0;       // if BTN1 is pressed

        if(Flash == 1)
        {
                digitalWrite(LD4, HIGH);
                delay(500);
                digitalWrite(LD4, LOW);
                delay(500);
        }
}
```

Figure 7.24: Program Listing of the Project

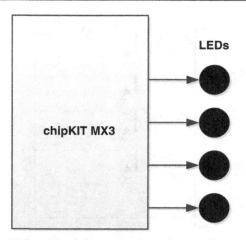

Figure 7.25: Block Diagram of the Project

7.7.3 Project Hardware

In this project, the PmodLED module is used. This is a small Pmod module with four LEDs as shown in Figure 7.26. The connection diagram of this module is shown in Figure 7.27. Each LED is driven from a transistor circuit with a minimum of 1 mA current.

The pin connections of the PmodLED module are as follows:

Pin Number	Pin Function
1	LD0
2	LD1
3	LD2
4	LD3
5	GND
6	VCC

Figure 7.26: The PmodLED Module

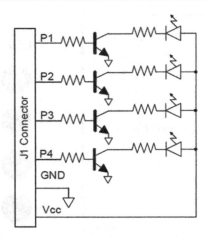

Figure 7.27: Connection Diagram of the PmodLED Module

When connected to Pmod connector JA, the interface between the module and the microcontroller I/O pins when the PmodLED is connected to the top row of the connector is as follows:

Pin Number	Microcontroller I/O Port Pin
1	RE0 (JA-01, logical I/O pin 0)
2	RE1 (JA-02, logical I/O pin 1)
3	RE2 (JA-03, logical I/O pin 2)
4	RE3 (JA-04, logical I/O pin 3)

Figure 7.28 shows the PmodLED module connected to the development board. The LEDs are configured as follows:

LD3 LD2 LD1 LD0
○   ○   ○   ○

7.7.4 Project PDL

The operation of the project is described in the PDL given in Figure 7.29. At the beginning of the program, LD1–LD4 (logical ports 0–3) are configured as outputs. The LEDs are turned ON and OFF in a rotating manner.

7.7.5 Project Program

The program is called ROTATE1, and the program listing is shown in Figure 7.30. At the beginning of the program, logical ports 0–3 corresponding to LEDs LD0–LD3 are configured as outputs and all the LEDs are turned OFF. Then, the LEDs are turned ON one at a time in a rotating manner with 1-s delay between each output.

Figure 7.28: Connecting the PmodLED Module to the Development Board

7.7.6 Modified Program

The program given in Figure 7.30 can be modified and made more efficient by using an array to store the port I/O logical port numbers. The modified program is called ROTATE2 and is shown in Figure 7.31. In this version of the program, integer array PmodLED is loaded with the logical port I/O numbers. These ports are then configured as outputs inside the *setup()* function. Inside the main program, a loop is formed to send data to the LEDs in a rotating manner.

7.7.7 Another Version of the Program

Another version of the program is given in this section where data is sent to PORTE where the LEDs are connected to in order to turn the LEDs ON and OFF. This version of the program is called ROTATE3 and is shown in Figure 7.32. PORTE is configured as output by

```
BEGIN
        Configure logical I/O pin 0 as output
        Configure logical I/O pin 1 as output
        Configure logical I/O pin 2 as output
        Configure logical I/O pin 3 as output
        Turn all LEDs OFF
        DO FOREVER
                Turn LD0 ON
                Wait 1 second
                Turn LD0 OFF
                Turn LD1 ON
                Wait 1 second
                Turn LD1 OFF
                Turn LD2 ON
                Wait 1 second
                Turn LD2 OFF
                Turn LD3 ON
                Wait 1 second
                Turn LD3 OFF
        ENDDO
END
```

Figure 7.29: PDL of the Project

clearing the TRISE register. Then, a variable called **j** is used in the main program loop to turn the LEDs ON and OFF. This variable takes the values of 1, 2, 4, 8, 1, 2, ...

7.8 Project 7.8 – Random Flashing LEDs

7.8.1 Project Description

In this project, four LEDs are used as in Project 7.7 and these LEDs are turned ON and OFF randomly with 1-s delay between each output.

7.8.2 Block Diagram

The block diagram of the project is as in Figure 7.25.

7.8.3 Project Hardware

In this project, the PmodLED module is used as in Project 7.7 and is connected to Pmod connector JA, that is, the four LEDs are connected to PORTE low nibble.

7.8.4 Project PDL

The operation of the project is described in the PDL given in Figure 7.33. At the beginning of the program, PORTE is configured as output. Then, inside the main program loop, a random number is generated between 1 and 15, and this number is sent to PORTE to turn the LEDs ON and OFF accordingly.

```
/*=======================================================================
                        ROTATING LEDs
                        ==============

This program uses a PmodLEd module with 4 LEDs.
The LEDs are turned ON and OFF in a rotating manner with one second delay between
each output.

The PmodLED module is connected to connector JA and thus the logical I/O port
numbers of the LEDs are 0, 1, 2, 3.

Author:         Dogan Ibrahim
Date:           May, 2014
File:           ROTATE1
Board:          chipKIT MX3
=======================================================================*/
int LD0 = 0;                            // LD0 logical port number
int LD1 = 1;                            // LD1 logical port number
int LD2 = 2;                            // LD2 logical port number
int LD3 = 3;                            // LD3 logical port number

void setup()
{
        pinMode(LD0, OUTPUT);           // set LD0 as output
        pinMode(LD1, OUTPUT);           // set LD1 as output
        pinMode(LD2, OUTPUT);           // set LD2 as output
        pinMode(LD3, OUTPUT);           // set LD3 as output
        digitalWrite(LD0, LOW);         // turn LD0 OFF to start with
        digitalWrite(LD1, LOW);         // turn LD1 OFF to start with
        digitalWrite(LD2, LOW);         // turn LD2 OFF to start with
        digitalWrite(LD3, LOW);         // turn LD3 OFF to start with
}


void loop()
{
        digitalWrite(LD0, HIGH);        // turn LD0 ON
        delay(1000);                    // wait 1 second
        digitalWrite(LD0, LOW);         // turn LD0 OFF
        digitalWrite(LD1, HIGH);        // turn LD1 ON
        delay(1000);                    // wait 1 second
        digitalWrite(LD1, LOW);         // turn LD1 OFF
        digitalWrite(LD2, HIGH);        // turn LD2 ON
        delay(1000);                    // wait 1 second
        digitalWrite(LD2, LOW);         // turn LD2 OFF
        digitalWrite(LD3, HIGH);        // turn LD3 ON
        delay(1000);                    // wait 1 second
        digitalWrite(LD3, LOW);         // turn LD3 OFF
}
```

Figure 7.30: Program Listing of the Project

```
/*============================================================================
                              ROTATING LEDs
                              =============

This program uses a PmodLEd module with 4 LEDs.
The LEDs are turned ON and OFF in a rotating manner with one second delay between
each output.

The PmodLED module is connected to connector JA and thus the logical I/O port
numbers of the LEDs are 0, 1, 2, 3.

In this version of the program an array is used to store the logical port numbers

Author:        Dogan Ibrahim
Date:          May, 2014
File:          ROTATE2
Board:         chipKIT MX3
=============================================================================*/
int PmodLED[] ={ 0, 1, 2, 3};                        // PmodLED logical port numbers

void setup()
{
        unsigned char j;
        for(j = 0; j <= 3; j++)                      // do for all LEDs
        {
                pinMode(PmodLED[j], OUTPUT);         // set LEDs as output
                digitalWrite(PmodLED[j], LOW);       // turn LEDs OFF to start with
        }
}


void loop()
{
        unsigned char j;
        for(j = 0; j <= 3; j++)                      // do for all LEDs
        {
                digitalWrite(PmodLED[j], HIGH);      // turn LED ON
                delay(1000);                         // wait 1 second
                digitalWrite(PmodLED[j], LOW);       // turn LD0 OFF
        }
}
```

Figure 7.31: Modified Program

7.8.5 Project Program

The program is called RANDOM, and the program listing is shown in Figure 7.34. At the
beginning of the program, register TRISE is cleared to make PORTE pins outputs. Then,
inside the main program, a random number is generated between 1 and 15 using the built-in

```
/*=========================================================================
                       ROTATING LEDs
                       =============

This program uses a PmodLEd module with 4 LEDs.
The LEDs are turned ON and OFF in a rotating manner with one second delay between
each output.

The PmodLED module is connected to connector JA and thus the logical I/O port
numbers of the LEDs are 0, 1, 2, 3.

In this version of the program an array is used to store the logical port numbers and 4-bits
are combined into a 4-bit port

Author:         Dogan Ibrahim
Date:           May, 2014
File:           ROTATE3
Board:          chipKIT MX3
=========================================================================*/
unsigned char j = 1;

void setup()
{
        TRISE = 0;                              // Configure PORTE as output
}



void loop()
{
        PORTE = j;                              // Output to PORTE
        delay(1000);                            // wait 1 second
        j = j << 1;                             // shift left
        if(j == 16) j = 1;                      // back to 1 if 16
}
```

Figure 7.32: Another Version of the Program

```
BEGIN
        Configure PORTE as output
        DO FOREVER
                Generate a random number between 1 and 15
                Send this random number to PORTE
                Wait 1 second
        ENDDO
END
```

Figure 7.33: PDL of the Project

```
/*========================================================================
                        RANDOM FLASHING LEDs
                        ====================

This program uses a PmodLEd module with 4 LEDs.
The LEDs are turned ON and OFF randomly.

The PmodLED module is connected to connector JA and thus they are connected to the
lower nibble of PORTE


Author:         Dogan Ibrahim
Date:           May, 2014
File:           RANDOM
Board:          chipKIT MX3
========================================================================*/
unsigned Ran = 10;

void setup()
{
        TRISE = 0;                              // Configure PORTE as output
}



void loop()
{
        Ran = random(1, 15);                    // generate a random number 1-15
        PORTE = Ran;                            // output to PORTE
        delay(1000);                            // wait 1 second
}
```

Figure 7.34: Program Listing of the Project

random function, and this number is sent to PORTE. This process is repeated continuously after 1-s delay.

7.9 Project 7.9 – Fading LED

7.9.1 Project Description

In this project, an external LED is connected to port pin RD0 (logical port number 22) of the microcontroller through a current-limiting resistor. RD0 is pin JC-09 of the Pmod connector JC. The program changes the brightness of the LED gradually from OFF to full brightness by sending analogue voltage in the form of PWM signal to the LED.

The program uses the built-in function **analogWrite(pin, duty)**, which sends a PWM signal to the specified pin with the specified duty cycle (0–255). The following I/O pins of the microcontroller can be used with the **analogWrite** function:

Pmod Connector	Logical Pin Number	Microcontroller I/O Pin
JC-09	22	RD0
JC-10	23	RD1
JD-02	25	RD2
JD-08	29	RD3
JB-09	14	RD4

In this project, I/O pin RD0 (logical pin number 22) is used.

7.9.2 Block Diagram

The block diagram of the project is shown in Figure 7.35.

7.9.3 Project Hardware

In this project, the PmodBB module is used. This board contains a small breadboard and Pmod connectors. An LED is placed on the breadboard and connected to pin JC-09 of the microcontroller through a 120 Ω current-limiting resistor and a short Pmod cable. The Pmod cable is connected to the bottom row of the Pmod connector. Pin 9 of connector JC is the third pin position of the bottom connector (bottom connector pins are 7 to 12 from left to right).

The PmodBB module with the LED and the resistor is shown in Figures 7.36 and 7.37. The circuit diagram of the project is shown in Figure 7.38.

7.9.4 Project PDL

The operation of the project is described in the PDL given in Figure 7.39. At the beginning of the program, logical port pin 22 is configured as output. Then, inside the main program loop, a PWM signal is sent to the port with varying duty cycle. As a result, the LED brightness increases gradually.

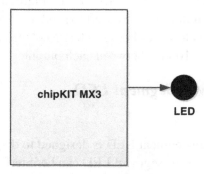

Figure 7.35: Block Diagram of the Project

Figure 7.36: The PmodBB Module and chipKIT MX3

7.9.5 Project Program

The program is called FADE, and the program listing is shown in Figure 7.40. At the beginning of the program, logical port 22 (port pin RD0) is configured as output. Then, inside the main program loop, function **analogWrite** is used to send a PWM signal to the port with varying duty cycle. The duty cycle varies between 0 (LED OFF) and 255 (full brightness) in steps of five and with a delay of 100 ms between each output.

7.10 Project 7.10 – Seven-Segment LED

7.10.1 Project Description

In this project, a two-digit seven-segment LED is designed to display the number "27." The project shows how a two-digit seven-segment LED can be interfaced and used with the chipKIT MX3 development board.

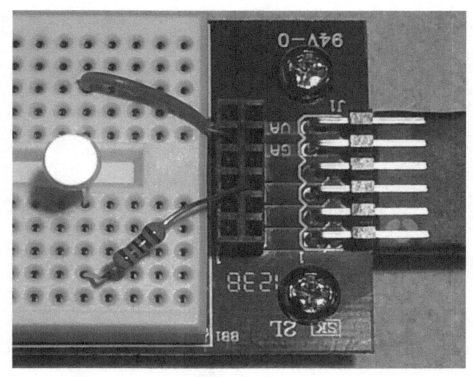

Figure 7.37: The PmodBB Module With LED and Resistor

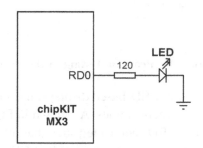

Figure 7.38: Circuit Diagram of the Project

```
BEGIN
        Configure logical port pin 22 as output
        DO FOREVER
                Send PWM signal to logical port 22 with increasing duty cycle
        ENDDO
END
```

Figure 7.39: PDL of the Project

```
/*=============================================================================
                        FADING LED
                        ==========

In this program an external LED is connected to port pin RD0 (logical port number 22)
of the microcontroller.

The program sends analog voltages (PWM) to the LED to change the brightness of the LED
from OFF to full brightness.


Author:         Dogan Ibrahim
Date:           May, 2014
File:           FADE
Board:          chipKIT MX3
===============================================================================*/
int LED = 22;                           // LED logical port number
unsigned char Brightness = 0;

void setup()
{
pinMode(LED, OUTPUT);                   // set LED as output
}


void loop()
{
        analogWrite(LED, Brightness);
        Brightness= Brightness + 5;
        if(Brightness > 255)Brightness = 0;
        delay(100);

}
```

Figure 7.40: Program Listing of the Project

Seven-segment displays are basically LED-based displays, used frequently in electronic
circuits to show numeric or alphanumeric values. As shown in Figure 7.41, a seven-segment
display basically consists of seven LEDs connected such that numbers from 0 to 9 and some
letters can be displayed when appropriate segments of the LEDs are turned ON. Segments are
identified by letters from **a** to **g**, and Figure 7.42 shows the segment names of a typical seven-
segment display.

Figure 7.43 shows how numbers from 0 to 9 can be obtained by turning ON different LED
segments of the display.

Seven-segment displays are available in two different configurations: **common cathode** and
common anode. As shown in Figure 7.44, in common cathode configuration, all the cath-
odes of all LED segments are connected together to ground. The segments are turned ON by

Figure 7.41: Four-Digit and One-Digit Seven-Segment Displays

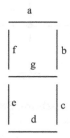

Figure 7.42: Segment Names of a Seven-Segment Display

applying a logic 1 to the required LED segment via current-limiting resistors. In common cathode configuration, the seven-segment LED is connected to the microcontroller in current-sourcing mode.

In a common anode configuration, the anode terminals of all the LEDs are connected together as shown in Figure 7.45. This common point is then normally connected to the supply voltage. A segment is turned ON by connecting its cathode terminal to logic 0 via a current-limiting resistor. In common anode configuration, the seven-segment LED is connected to the microcontroller in current-sinking mode.

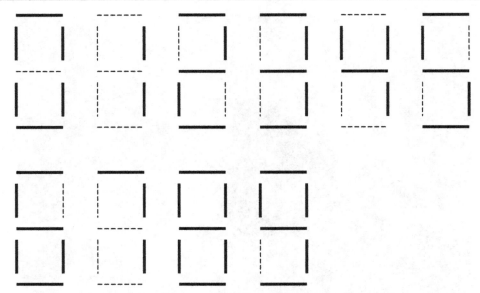

Figure 7.43: Displaying Numbers 0–9

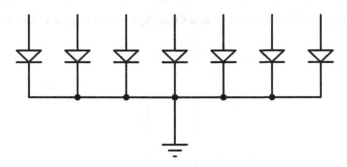

Figure 7.44: Common Cathode Seven-Segment Display

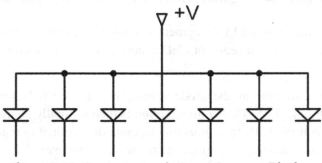

Figure 7.45: Common Anode Seven-Segment Display

In seven-segment display applications, a table is normally constructed that shows the relationship between the numbers to be displayed and the LED segments that should be turned ON to display the required number. In single-digit applications (to display 0–9), the common pin is connected to ground or to +V supply depending on whether the display is of type common cathode or common anode.

In multidigit seven-segment display applications (see Figure 7.46), digit pins are tied in parallel except the common pins. The common pin of each digit is turned ON alternately by the microcontroller. By displaying each digit for several milliseconds, the eye cannot differentiate that the digits are not ON all the time. This way we can multiplex any number of seven-segment displays together. For example, to display number 45, we have to send 4 to the first digit and enable its common pin. After a few milliseconds, number 5 is sent to the second digit and the common point of the second digit is enabled. When this process is repeated continuously, the user sees as if both displays are ON continuously.

Some manufacturers provide multiplexed multidigit displays in single packages. For example, we can purchase two-, four-, or eight-digit multiplexed displays in a single package. As an example, a two-digit display can be controlled from the microcontroller as follows:

- Send the segment bit pattern for digit 1 to segments a–g.
- Enable digit 1.
- Wait for a few milliseconds.
- Disable digit 1.
- Send the segment bit pattern for digit 2 to segments a–g.
- Enable digit 2.
- Wait for a few milliseconds.
- Disable digit 2.
- Repeat the above process continuously.

In this project, a two-digit, seven-segment Pmod-compatible display module called PmodSSD is used. The block diagram of the project is shown in Figure 7.47.

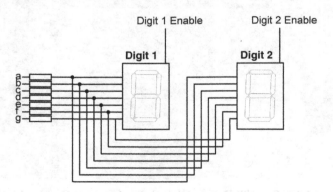

Figure 7.46: Multiplexed Two-Digit Seven-Segment Displays

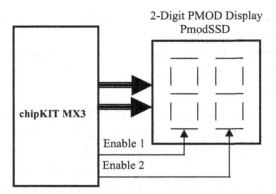

Figure 7.47: Block Diagram of the Project

7.10.2 Project Hardware

As shown in Figure 7.48, PmodSSD module is a high-brightness two-digit, common cathode seven-segment display with built-in current-limiting resistors. Figure 7.49 shows the connection diagram of the display.

The PmodSSD module has 2×6 pins, and in this project the module is connected to PMOD connectors JA and JB. Therefore, the relationship between the port numbers, port names, and

Figure 7.48: PmodSSD Two-Digit Seven-Segment Display Module

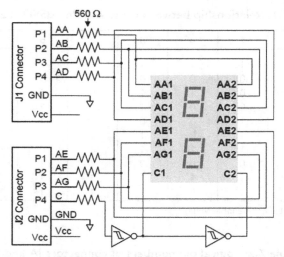

Figure 7.49: Connection Diagram of PmodSSD Module

PmodSSD segment are as in Table 7.1. Table 7.2 shows the logical port numbers of the JA and JB connectors. Notice that the digits are enabled from bit 3 of connector JB. The logical port numbers and the bits to be sent to these port pins are shown in Table 7.3 where the horizontal numbers from 0 to 9 are the numbers to be displayed on a digit and numbers on the left-hand side (0, 1, 2, 3, 8, 9, 10) are the logical port numbers. As an example, to display number 2 on the MSD position, we have to send the following bits to the logical ports:

Logical Port Number	Data to Be Sent
0	1
1	1
2	0
3	1
8	1
9	0
10	1

In addition, logical port number 11 must be set HIGH to enable the MSD.

The interface between the chipKIT MX3 and the display is shown in Figure 7.50. In this figure, the logical I/O port numbers are shown.

7.10.3 Project PDL

The operation of the project is described in the PDL given in Figure 7.51. At the beginning of the program, the logical ports used in the project are configured as outputs. Then, number 2

Table 7.1: Relationship between ports and PmodSSD segments.

Number	x	g	f	e	d	c	b	a	JB	JA
0	c	0	1	1	1	1	1	1	0x3	0xF
1	c	0	0	0	0	1	1	0	0x0	0x6
2	c	1	0	1	1	0	1	1	0x5	0xB
3	c	1	0	0	1	1	1	1	0x4	0xF
4	c	1	1	0	0	1	1	0	0x6	0x6
5	c	1	1	0	1	1	0	1	0x6	0xD
6	c	1	1	1	1	1	0	1	0x7	0xD
7	c	0	0	0	0	1	1	1	0x0	0x7
8	c	1	1	1	1	1	1	1	0x7	0xF
9	c	1	1	0	1	1	1	1	0x6	0xF

c is the control bit (0 = enable LSD; 1 = enable MSD).

Table 7.2: Logical pin numbers of connectors JA and JB.

JB-04	JB-03	JB-02	JB-01	JA-04	JA-03	JA-02	JA-01
11	10	9	8	3	2	1	0
RF06	RF02	RF03	RD09	RE03	RE02	RE01	RE00

Table 7.3: Logical port numbers and numbers to be displayed.

	0	1	2	3	4	5	6	7	8	9
0	1	0	1	1	0	1	1	1	1	1
1	1	1	1	1	1	0	0	1	1	1
2	1	1	0	1	1	1	1	1	1	1
3	1	0	1	1	0	1	1	0	1	1
8	1	0	1	0	0	0	1	0	1	0
9	1	0	0	0	1	1	1	0	1	1
10	0	0	1	1	1	1	1	0	1	1

is sent to the display and the MSD is enabled. After a delay of 5 ms, number 7 is sent and the LSD is enabled. The display remains in this state for a further 5 ms. This process is repeated forever, thus displaying number 27.

7.10.4 Project Program

The program is called SSD1, and the program listing is shown in Figure 7.52. At the beginning of the program, logical port pins 0, 1, 2, 3, 8, 9, 10, and 11 are configured as outputs. An array called PmodSSD is created to store the logical pin numbers. Similarly, a two-dimensional array called **segment** is created to store the bit pattern for each number to be displayed. The program creates a loop for each digit, where inside these loops the logical ports where

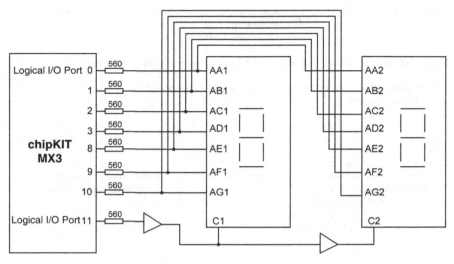

Figure 7.50: Display Interface

BEGIN
 Configure logical port pins 0,1,2,3,8,9,10,11 as outputs
 DO FOREVER
  Send bit pattern for number 2 to the display
  Enable MSD digit
  Wait 5 milliseconds
  Send bit pattern for number 7 to the display
  Disable MSD digit
  Enable LSD digit
  Wait 5 milliseconds
  Disable MSD digit
 ENDDO
END

Figure 7.51: PDL of the Project

the display is connected to are selected and the correct column of array **segment** is accessed (indexed by the number to be displayed, 0–9) and the bits in this column are sent to the display so that the required number can be displayed.

Figure 7.53 shows the PmodSSD module connected to the chipKIT MX3 development board.

7.11 Project 7.11 – Seven-Segment Single-Digit LED Counter

7.11.1 Project Description

In this project, one digit of the two-digit PmodSSD module is used to count up from 0 to 9 with 1-s delay between each count.

```
/*=======================================================================
                       2-DIGIT 7-SEGMENT DISPLAY
                       =========================

This program uses a PmodSSD module with 2-digit 7-segment display.

The module is connected to Pmod connectors JA and JPB. Number 27 is
displayed on the display

Author:        Dogan Ibrahim
Date:          May, 2014
File:          SSD1
Board:         chipKIT MX3
=======================================================================*/
int PmodSSD[] ={0, 1, 2, 3, 8, 9, 10, 11};              // PmodSSD logical port numbers
unsigned char segment[7][10] =
                {1,0,1,1,0,1,1,1,1,1,
                 1,1,1,1,1,0,0,1,1,1,
                 1,1,0,1,1,1,1,1,1,1,
                 1,0,1,1,0,1,1,0,1,1,
                 1,0,1,0,0,0,1,0,1,0,
                 1,0,0,0,1,1,1,0,1,1,
                 0,0,1,1,1,1,1,0,1,1};

void setup()
{
        unsigned char j;
        for(j = 0; j <= 7; j++)                         // do for all required ports
        {
                pinMode(PmodSSD[j], OUTPUT);            // set ports as outputs
        }
}

void loop()
{
    unsigned char j, msd, lsd;

    msd = 2;                                            // msd of number to be displayed
    lsd = 7;                                            // lsd of number to be displayed

      for(;;)                                           // do forever
      {
              for(j = 0; j < 7;j++)                     // do for all 7 bits
              {
                    digitalWrite(PmodSSD[j], segment[j][msd]);
              }
              digitalWrite(11, HIGH);                   // enable MSD digit
              delay(5);                                 // wait 5 ms

              for(j = 0; j < 7;j++)                     // do for all 7 bits
              {
                    digitalWrite(PmodSSD[j], segment[j][lsd]);
              }
              digitalWrite(11, LOW);                    // enable LSD digit
              delay(5);                                 // wait 5 ms
      }
}
```

Figure 7.52: Program Listing of the Project

Figure 7.53: PmodSSD Module Displaying the Required Number

7.11.2 Project Hardware

In this project, the PmodSSD is connected to connectors JA and JB as in Project 7.10 and only the MSD is used.

7.11.3 Project PDL

The operation of the project is described in the PDL given in Figure 7.54. At the beginning of the program, the logical ports used in the project are configured as outputs. Then, the value of a variable that is incremented every second is sent to the MSD of the display. Thus, the display counts as 0, 1, 2, 3, 4, 5, 6, 7, 8, 9, 0, 1, ...

```
BEGIN
        Configure logical port pins 0,1,2,3,8,9,10,11 as outputs
        Initialize variable Cnt to zero
        DO FOREVER
                Send Cnt to MSD
                Enable MSD digit
                Wait 1 second
                Increment Cnt
        ENDDO
END
```

Figure 7.54: PDL of the Project

7.11.4 Project Program

The program is called SSD2, and the program listing is shown in Figure 7.55. At the beginning of the program, logical port pins 0, 1, 2, 3, 8, 9, 10, and 11 are configured as outputs. An array called PmodSSD is created to store the logical pin numbers. Similarly, an array called **segment** is created to store the bit pattern for the number to be displayed. Variable Cnt is used as the counter. The program creates a loop for the digit, where inside the loop the logical ports where the display is connected to are selected and the correct column of array **segment** is accessed (indexed by variable Cnt) and the bits in this column are sent to the display so that the value of **Cnt** can be displayed.

7.12 Project 7.12 – Using LCD Display

7.12.1 Project Description

In this project, the use of an LCD display is described. The project displays the text "chipKIT MX3" on the first row of an LCD.

In microcontroller systems, the output of a measured variable is usually displayed using LEDs, seven-segment displays, or LCD-type displays. LCDs have the advantages that they can be used to display alphanumeric or graphical data. Some LCDs have 40 or more character lengths with the capability to display several lines. Some other LCD displays can be used to display graphics images. Some modules offer colour displays while some others incorporate backlighting so that they can be viewed in dimly lit conditions.

There are basically two types of LCDs as far as the interface technique is concerned: parallel LCDs and serial LCDs. Parallel LCDs (e.g., Hitachi HD44780) are connected to a microcontroller using more than one data line and the data is transferred in parallel form. It is common to use either four or eight data lines. Using a four-wire connection saves I/O pins, but it is slower since the data is transferred in two stages. Serial LCDs are connected to the microcontroller using only one data line, and data is usually sent to the LCD using the standard RS-232 asynchronous data communication protocol. Serial LCDs are much easier to use, but they cost more than the parallel ones.

The programming of a parallel LCD is usually a complex task and requires a good understanding of the internal operation of the LCD controllers, including the timing diagrams. Fortunately, most high-level languages provide special library commands for displaying data on alphanumeric as well as on graphical LCDs. All the user has to do is connect the LCD to the microcontroller, define the LCD connection in the software, and then send special commands to display data on the LCD.

7.12.2 HD44780 LCD Module

HD44780 is one of the most popular alphanumeric LCD modules used in industry and also by hobbyists. This module is monochrome and comes in different sizes. Modules with 8, 16, 20,

```
/*========================================================================
                        1-DIGIT 7-SEGMENT UP COUNTER
                        ==============================

This program uses the MSD digit of thePmodSSD module to count up with one second
Delay between each output.

The module is connected to Pmod connectors JA and JPB

Author:         Dogan Ibrahim
Date:           May, 2014
File:           SSD2
Board:          chipKIT MX3
=========================================================================*/
int PmodSSD[] ={0, 1, 2, 3, 8, 9, 10, 11};               // PmodSSD logical port numbers
unsigned char segment[7][10] =
                {1,0,1,1,0,1,1,1,1,1,
                 1,1,1,1,1,0,0,1,1,1,
                 1,1,0,1,1,1,1,1,1,1,
                 1,0,1,1,0,1,1,0,1,1,
                 1,0,1,0,0,0,1,0,1,0,
                 1,0,0,0,1,1,1,0,1,1,
                 0,0,1,1,1,1,1,0,1,1};

unsigned char j, Cnt = 0;

void setup()
{
        unsigned char j;
        for(j = 0; j <= 7; j++)                          // do for all required ports
        {
                pinMode(PmodSSD[j], OUTPUT);             // set ports as outputs
        }
        digitalWrite(11, HIGH);                          // enable MSD digit

}


void loop()
{
                for(j = 0; j < 7;j++)                    // do for all 7 bits
                {
                        digitalWrite(PmodSSD[j], segment[j][Cnt]);
                }

                delay(1000);                             // wait 1 second
                Cnt++;                                   // increment Cnt
                If(Cnt == 10)Cnt = 0;                    // reset Cnt
}
```

Figure 7.55: Program Listing of the Project

Table 7.4: Pin configuration of HD44780 LCD module.

Pin No.	Name	Function
1	V_{SS}	Ground
2	V_{DD}	Positive supply
3	V_{EE}	Contrast
4	RS	Register select
5	R/W	Read/write
6	E	Enable
7	D0	Data bit 0
8	D1	Data bit 1
9	D2	Data bit 2
10	D3	Data bit 3
11	D4	Data bit 4
12	D5	Data bit 5
13	D6	Data bit 6
14	D7	Data bit 7

24, 32, and 40 columns are available. Depending on the model chosen, the number of rows varies between 1, 2, and 4. The display provides a 14-pin (or 16-pin) connector to a microcontroller. Table 7.4 gives the pin configuration and pin functions of a 14-pin LCD module. Below is a summary of the pin functions.

V_{SS} is the 0 V supply or ground. The V_{DD} pin should be connected to the positive supply. Although the manufacturers specify a 5 V D.C. supply, the modules will usually work with as low as 3 V or as high as 6 V.

Pin 3 is named V_{EE}, and this is the contrast control pin. This pin is used to adjust the contrast of the display, and it should be connected to a variable voltage supply. A potentiometer is normally connected between the power supply lines with its wiper arm connected to this pin so that the contrast can be adjusted.

Pin 4 is the register select (RS), and when this pin is LOW, data transferred to the display is treated as commands. When RS is HIGH, character data can be transferred to and from the module.

Pin 5 is the read/write (R/W) line. This pin is pulled LOW in order to write commands or character data to the LCD module. When this pin is HIGH, character data or status information can be read from the module.

Pin 6 is the enable (E) pin that is used to initiate the transfer of commands or data between the module and the microcontroller. When writing to the display, data is transferred only on the HIGH-to-LOW transition of this line. When reading from the display, data becomes available after the LOW-to-HIGH transition of the enable pin, and this data remains valid as long as the enable pin is at logic HIGH.

Pins 7–14 are the eight data bus lines (D0–D7). Data can be transferred between the microcontroller and the LCD module either using a single 8-bit byte or as two 4-bit nibbles. In the latter case, only the upper four data lines (D4–D7) are used. 4-Bit mode has the advantage that four less I/O lines are required to communicate with the LCD.

7.12.3 Connecting the LCD to the Microcontroller

In this book, the Pmod LCD module called PmodCLP (Pmod Character Parallel LCD) is used in 8-bit mode. As shown in Figure 7.56, this module has two Pmod connectors: a dual connector called J1 and a single connector called J2 (see Figure 7.57). The module contains a 3.3–5 V converter for the LCD. The interface contains eight data signals and three control signals. The three control signals are the R/W, RS, and E.

In this project, the PmodCLP module is connected to Pmod connectors JA and JB. The interface between the LCD module and the Pmod connectors is shown in Table 7.5. The brightness of the display is set by a resistive potential divider consisting of a 100 Ω and a 2.1k resistor.

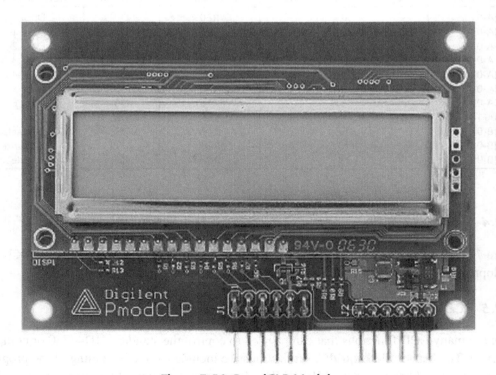

Figure 7.56: PmodCLP Module

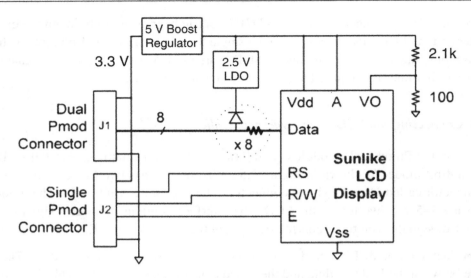

Figure 7.57: PmodCLP Connection Diagram

Table 7.5: PmodCLP interface.

Connector Pin	Logical I/O Port Number	Port Name	LCD Pin	LCD Pin Name
JA-01	0	RE0	D0	Data 0
JA-02	1	RE1	D1	Data 1
JA-03	2	RE2	D2	Data 2
JA-04	3	RE3	D3	Data 3
JA-07	4	RE4	D4	Data 4
JA-08	5	RE5	D5	Data 5
JA-09	6	RE6	D6	Data 6
JA-10	7	RE7	D7	Data 7
JB-07	12	RD6	RS	Register select
JB-08	13	RD5	R/W	Read/write
JB-09	14	RD4	E	Enable

7.12.4 Project Hardware

Figure 7.58 shows the connection between the PmodCLP module and the chipKIT MX3 development board. Here, the logical I/O port numbers are shown.

7.12.5 LCD Functions

There are many LCD functions that can be used to control the standard HD44780 or compatible LCD. The header file **LiquidCrystal.h** must be included at the beginning of the program by *importing* this library into the MPIDE IDE. Some commonly used functions are described in this section briefly (in these functions, variable name **lcd** is used).

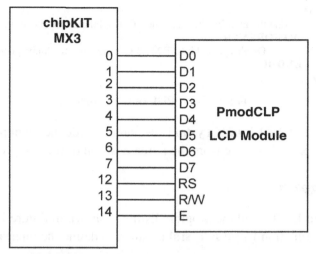

Figure 7.58: PmodCLP Interface to the Development Board

LiquidCrystal lcd(rs, rw, e, d0, d1, d2, d3, d4, d5, d6, d7): This function creates a variable called **lcd** of type **LiquidCrystal** and is used to define the interface between the microcontroller logical I/O port numbers and the LCD.

lcd.begin(col, row): This function must be called before any other LCD function, and it initialises the LCD size by specifying its number of rows and columns.

lcd.clear(): This function clears the LCD screen and positions the cursor in the top left-hand corner.

lcd.home(): This function homes the cursor by positioning it in the top left-hand corner.

lcd.setCursor(col, row): This function positions the cursor at the specified column and row position.

lcd.write(byte): This function writes a character (or byte) at the current cursor position.

lcd.print(text): This function writes the *text* at the current cursor position.

lcd.print(data, BIN|DEC|OCT|HEX): This function writes the data in the specified number base. The data can be byte, integer, long, or string.

lcd.blink() and lcd.noBlink(): These functions start or stop the cursor from blinking.

lcd.cursor and lcd.noCursor(): These functions enable or disable the display of the cursor.

lcd.scrollDisplayLeft() and lcd.scrollDisplayRight(): These functions scroll the display one position to the left or right.

7.12.6 Project PDL

The operation of the project is described in the PDL given in Figure 7.59. At the beginning of the program, the interface between the LCD and the chipKIT MX3 board is specified.

BEGIN
Define the interface between the LCD and the development board
DO FOREVER
Display text chipKIT MX3 starting from the home position
ENDDO
END

Figure 7.59: PDL of the Project

A variable called **lcd** of type **LiquidCrystal** is declared. Inside the main program, the text **chipKIT MX3** is displayed starting from the home position of the display.

7.12.7 Project Program

The program is called LCD1, and the program listing is shown in Figure 7.60. At the beginning of the program, function **LiquidCrystal** is called to define the interface between the

```
/*=======================================================================
                    LCD DISPLAY
                    ==========

This program uses the PmodCLP LCD module. The module is connected to Pmod
Connectors JA and JPB.

The program displays the text "chipKIT MX3" on the first row of the LCD.

Author:       Dogan Ibrahim
Date:         May, 2014
File:         LCD1
Board:        chipKIT MX3
=======================================================================*/

#include <LiquidCrystal.h>

LiquidCrystallcd(12, 13, 14, 0, 1, 2, 3, 4, 5, 6, 7);


void setup()
{
      lcd.begin(16, 2);                              // LCD is 16 columns, 2 rows
}


void loop()
{
      lcd.print("chipKIT MX3");                      // display text
      while(1);                                      // stop here
}
```

Figure 7.60: Program Listing of the Project

LCD and the logical I/O ports. Then, the text **chipKIT MX3** is displayed starting from the home position of the cursor.

Figure 7.61 shows the PmodCLP module connected to the development board and displaying the required text.

Figure 7.61: Displaying Text chipKIT MX3

7.13 Project 7.13 – Scrolling LCD Display

7.13.1 Project Description

In this project, it is shown how text displayed on the LCD can easily be scrolled left and right. The text **chipKIT MX3** is initially displayed on the LCD. Then, this text is scrolled to the end of the screen towards the right and then towards the left of the display.

7.13.2 Project Hardware

In this project, the PmodCLP LCD module is used as in the previous project. The connection of the PmodCLP module to the chipKIT MX3 development board is as shown in Figure 7.58.

7.13.3 Project PDL

The operation of the project is described in the PDL shown in Figure 7.62. At the beginning of the program, the interface between the LCD and the chipKIT MX3 board is specified. A variable called **lcd** of type **LiquidCrystal** is declared. Inside the main program, the text **chipKIT MX3** is initially displayed at the top row of the LCD. Then, the main program loop starts where the text is initially scrolled to the right to the end of the screen (16 positions). After this, the text is scrolled to the left (27 positions) and then to the right (27 positions) continuously. A small delay (200 ms) is used between each output.

7.13.4 Project Program

The program is called LCD2, and the program listing is shown in Figure 7.63. At the beginning of the program, function **LiquidCrystal** is called to define the interface between the LCD and the logical I/O ports. Then, the text **chipKIT MX3** is displayed starting from the home position of the cursor. Inside the main program, the text is scrolled to the right end of the LCD screen (by either 16 or 27 positions depending on whether this is the first iteration or not). Then, the text is scrolled to the left end of the screen by 27 positions. The program waits 200 ms before each output.

```
BEGIN
        Define the interface between the LCD and the development board
        Display text chipKIT MX3
        DO FOREVER
                IF first time THEN
                        Scroll right by 16 positions
                ELSE
                        Scroll right by 27 positions
                ENDIF
                Scroll left 27 position
        ENDDO
END
```

Figure 7.62: PDL of the Project

```
/*=====================================================================
                        LCD SCROLLING EXAMPLE
                        ====================

    In this project, an LCD is connected to the chipKIT MX3 development board as in the
    previous project.

    The text "chipKIT MX3" is scrolled left and right. The LCD functions scrollDisplayLeft()
    andscrollDisplayRight() are used in the program


        Author:         Dogan Ibrahim
        Date:           May, 2014
        File:           LCD2
        Board:          chipKIT MX3
    =================================================================*/
    int cursor;
    int cursor_end;
    int flag = 0;
    //
    // Include the LiquidCrystal library in the program
    //
    #include <LiquidCrystal.h>

    //
    // Initialize the LCD library with the numbers of the interface pins
    //
    LiquidCrystallcd(12, 13, 14, 0, 1, 2, 3, 4, 5, 6, 7);            // define LCD interface

    void setup()
    {
        lcd.begin(16, 2);                                           // 16 column,2 row
        lcd.print("chipKIT MX3");                                   // display text
        delay(1000);                                                // wait 1 second
    }

    void loop()
    {
            if(flag == 0)                                           // If first time...
            {
                    flag = 1;
                    cursor_end =16;
            }
            else
                    cursor_end = 27;
    //
    // Scroll by "cursor_end" locations to the right (offscreen)
    //
            for (cursor = 0; cursor <cursor_end; cursor++)
            {
                    lcd.scrollDisplayRight();                       // scroll right
                    delay(200);                                     // 200ms delay
            }
    //
    // scroll 27 positions (text length+string length) to the left (offscreen)
    //
            for (cursor = 0; cursor < 27; cursor++)
            {
                    lcd.scrollDisplayLeft();                        // scroll left
                    delay(200);                                     // 200ms delay
            }
    }
```

Figure 7.63: Program Listing of the Project

7.14 Project 7.14 – Seconds Counter With LCD Display

7.14.1 Project Description

In this project, the PmodCLP LCD module is connected to the chipKIT MX3 board as in the previous project. Here, the LCD counts up in seconds where the display is in the following format:

Count = nn

This project shows how the mixture of both text and numeric data can be displayed on the LCD.

7.14.2 Project Hardware

In this project, the PmodCLP LCD module is used as in the previous project. The connection of the PmodCLP module to the chipKIT MX3 development board is as shown in Figure 7.58.

7.14.3 Project PDL

The operation of the project is described in the PDL shown in Figure 7.64. At the beginning of the program, the interface between the LCD and the chipKIT MX3 board is specified. A variable called **lcd** of type **LiquidCrystal** is declared. An integer counter variable **cnt** is declared. Inside the main program, the value of this variable is displayed after the text **Count =**. The variable is incremented by 1, and the process is repeated after 1-s delay. Notice that in this program, the built-in function delay is used to create 1-s delay and as a result of this the timing is not accurate.

7.14.4 Project Program

The program is called LCD3, and the program listing is shown in Figure 7.65. At the beginning of the program, function **LiquidCrystal** is called to define the interface between the LCD module and the development board. Integer variable **cnt** is declared and set to 0. Inside the main program loop, the cursor is set to home position and the value of variable **cnt** is

```
BEGIN
        Define the interface between the LCD and the development board
        Declare integer variable called cnt
        DO FOREVER
                Display value of cnt after text Count =
                Increment variable cnt
                Wait one second
        ENDDO
END
```

Figure 7.64: PDL of the Project

```
*=========================================================================
                    LCD SECONDS COUNTER
                    ===================
```

In this project, an LCD is connected to the chipKIT MX3 development board as in the previous project.

The project counts up every second and the count is displayed on the LCD in the following Format:

```
            Count = nn
```

```
Author:         Dogan Ibrahim
Date:           May, 2014
File:           LCD3
Board:          chipKIT MX3
=========================================================================*/
int cnt = 0;

//
// Include the LiquidCrystal library in the program
//
#include <LiquidCrystal.h>

//
// Initialize the LCD library with the numbers of the interface pins
//
  LiquidCrystallcd(12, 13, 14, 0, 1, 2, 3, 4, 5, 6, 7);              // define LCD interface

void setup()
{
        lcd.begin(16, 2);                                           // 16 column,2 row
}


void loop()
{
        lcd.home();                                                 // home the cursor
        lcd.print("Count = ");                                      // display "Count = "
        lcd.print(cnt);                                             // display the Cnt in decimal
        cnt++;                                                      // increment count
        delay(1000);                                                // wait 1 second
}
```

Figure 7.65: Program Listing of the Project

Figure 7.66: The LCD Display

displayed after the text **Count** = . Variable **cnt** is then incremented by 1, and this process is repeated after 1-s delay.

Figure 7.66 shows the LCD display as the program is running.

7.15 Project 7.15 – Event Counter With LCD Display

7.15.1 Project Description

In this project, the PmodCLP LCD module is connected to the chipKIT MX3 board as in the previous project. Here, the LCD counts external events and displays the number of events on the LCD in the following format:

    Event = nn

External events are said to occur when microcontroller pin RB08 (Pmod connector JPC pin JC-01, logical I/O port number 16) goes from logic 0 to logic 1. In this project, the PmodBTN button module is connected to Pmod connector JC and button BTN 0 is used to simulate an external event.

7.15.2 Project Hardware

In this project, the PmodCLP LCD module is connected to Pmod connectors JA and JB. The PmodBTN module is connected to the top row of Pmod connector JC. Thus, BTN0 is effectively connected to logical port number 16 (connector JC-01).

The connections of the PmodCLP and the PmodBTN modules to the chipKIT MX3 development board are as shown in Figure 7.67.

7.15.3 Project PDL

The operation of the project is described in the PDL shown in Figure 7.68. At the beginning of the program, the interface between the LCD and the chipKIT MX3 board is specified. A variable called **lcd** of type **LiquidCrystal** is declared. An integer counter variable **cnt** is then declared and set to 0. Inside the main program, the push-button BTN0 is checked continuously, and when the button is pressed (BTN0 going from 0 to 1) and released (BTN going from 1 to 0), the value of **cnt** is incremented by 1 and its value is sent to the LCD. Thus, the LCD displays the number of external events occurring.

7.15.4 Project Program

The program is called LCD4, and the program listing is shown in Figure 7.69. At the beginning of the program, function **LiquidCrystal** is called to define the interface between the

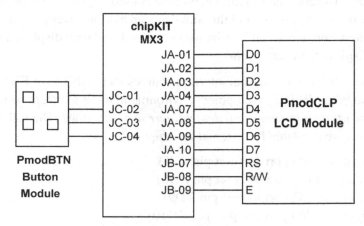

Figure 7.67: Connecting the PmodCLP and PmodBTN to the Development Board

```
BEGIN
        Define the interface between the LCD and the development board
        Configure BTN0 port as input
        Display Events = 0 to start with
        Set cnt = 0
        DO FOREVER
                Home the cursor
                Wait until button BTN0 is pressed
                Wait until button BTN0 is released
                Increment variable cnt by 1
                Display cnt
        ENDDO
END
```

Figure 7.68: PDL of the Project

LCD module and the development board. Variable **cnt** is declared and is set to 0. Also, BTN0 is assigned to logical I/O port number 16. Inside the *setup()* function, the LCD is configured as 2 rows, 16 columns, BTN0 is configured as input port, the cursor is set to home position, and text **Events = 0** is displayed to start with. Inside the main program loop, the state of BTN0 is checked continuously, and when the button is pressed (and then released), the value of **cnt** is incremented by 1 and its value is displayed on the LCD. Thus, the LCD displays the number of external events occurring.

Figure 7.70 shows the project hardware setup.

7.16 Project 7.16 – External Interrupt–Driven Event Counter With LCD Display

7.16.1 Project Description

In this project, the PmodCLP LCD module is connected to Pmod connectors JA and JB as in the previous few projects. The PmodBTN module is used to generate an external interrupt when button BTN0 is pressed, and this simulates an external event. Inside the interrupt service routine, the value of a variable is incremented by 1 and then displayed on the LCD. Thus, the LCD displays the event count.

PIC32 microcontroller has five external interrupt sources INT0–INT4 (or EXT_INT0 to EXT_INT4). Using the MPIDE development environment, these interrupt sources are accessed using the *attachInterrupt()* and *detachInterrupt()* functions. The following Pmod connector positions are available for external interrupts:

INT0: JB-04, logical I/O port 11 (port pin RF6)
INT1: JE-07, logical I/O port 36 (port pin RD8)
INT2: JB-01, logical I/O port 8 (port pin RD9)
INT3: JD-03, logical I/O port 26 (port pin RD10)
INT4: JD-09, logical I/O port 39 (port pin RD11)

```
/*=============================================================================
                            LCD EVENT COUNTER
                            =================
```

In this project, an LCD is connected to the chipKIT MX3 development board as in the previous project.i.e. to Pmod connectors JA and JPB.

In addition the PmodBTN 4-button module is connected to the top row of Pmod connector JPC.

The project counts and displays external events. An external event Is said to occur when button BTN0 is pressed.

The events are displayed in the following format:

            Events = nn


```
Author:         Dogan Ibrahim
Date:           May, 2014
File:           LCD4
Board:          chipKIT MX3
=============================================================================*/
int cnt = 0;                                        // initialize cnt to 0
int BTN0 = 16;                                       // logical I/O port for BTN0

//
// Include the LiquidCrystal library in the program
//
#include <LiquidCrystal.h>

//
// Initialize the LCD library with the numbers of the interface pins
//
  LiquidCrystallcd(12, 13, 14, 0, 1, 2, 3, 4, 5, 6, 7);        // define LCD interface

void setup()
{
        lcd.begin(16, 2);                            // 16 column,2 row
        pinMode(BTN0, INPUT);                        // set BTN0 as input
        lcd.home();
        lcd.print("Events = 0");
}


void loop()
{
        lcd.home();                                  // home the cursor
        while(digitalRead(BTN0) == 0);               // wait until button is pressed
        while(digitalRead(BTN0) == 1);               // wait until button is released
        cnt++;                                       // increment cnt
        lcd.print("Events = ");                      // display "Count = "
        lcd.print(cnt);                              // display the Cnt in decimal
}
```

Figure 7.69: Program Listing of the Project

Figure 7.70: Project Hardware Setup

In this project, a PmodBTN four-button Pmod module is connected to the bottom row of Pmod connector JE. Thus, BTN0 corresponds to logical port number 36, that is, interrupt number 1 (INT1).

7.16.2 Project Hardware

The PmodCLP LCD module is connected to Pmod connectors JA and JB. The PmodBTN module is connected to the bottom row of Pmod connector JPE. Thus, BTN0 is effectively connected to logical port number 36 (connector JE-07).

The connections of the PmodCLP and the PmodBTN modules to the chipKIT MX3 development board are as shown in Figure 7.71.

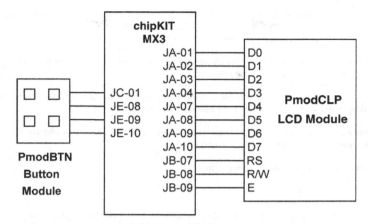

Figure 7.71: Connecting the PmodCLP and PmodBTN to the Development Board

7.16.3 Project PDL

The operation of the project is described in the PDL shown in Figure 7.72. At the beginning of the program, the interface between the LCD and the chipKIT MX3 board is specified. A variable called **lcd** of type **LiquidCrystal** is declared. The interrupt service routine is then configured. A variable is incremented inside the interrupt service routine, and its value is sent to the display to show the event count.

7.16.4 Project Program

The program is called INT, and the program listing is shown in Figure 7.73. At the beginning of the program, function **LiquidCrystal** is called to define the interface between the LCD module

```
BEGIN
        Define the interface between the LCD and the development board
        Display Events = 0 to start with
        Set cnt = 0
        Configure external interrupt INT1 with routine ISR
        DO FOREVER
        ENDDO
END


BEGIN/ISR
        Home the cursor
        Increment variable cnt by 1
        Display cnt
END/ISR
```

Figure 7.72: PDL of the Project

```
/*==============================================================================
                       INTERRUPT BASED LCD EVENT COUNTER
                       ================================

In this project, an LCD is connected to the chipKIT MX3 development board as in the previous
project. i.e. to Pmod connectors JA and JPB.

In addition the PmodBTN 4-button module is connected to the bottom row of Pmod connector JPE.
Thus, BTN0 button corresponds to pin JE-07 (interrupt number 1).

The project counts and displays external events. An external event is said to occur when button
BTN0 is pressed (interrupt pin goes from logic 0 to 1)

The events are displayed in the following format:

                       Events = nn

The program uses function attachInterrupt() to configure INT1 externalinterrupt (interrupt
number 1) such that when the interrupt pin (BTN0) goesfrom logic 0 to logic 1 (i.e. RISING)
then an interrupt is generated and theprogram jumps to interrupt service routine called ISR.

Author:        Dogan Ibrahim
Date:          May, 2014
File:          INT
Board:         chipKIT MX3
==============================================================================*/
int cnt = 0;                                             // initialize cnt to 0
//
// Include the LiquidCrystal library in the program
//
#include <LiquidCrystal.h>

//
// Initialize the LCD library with the numbers of the interface pins
//
  LiquidCrystallcd(12, 13, 14, 0, 1, 2, 3, 4, 5, 6, 7);  // define LCD interface

void setup()
{
        lcd.begin(16, 2);                                // 16 column,2 row
        lcd.home();                                      // home the cursor
        lcd.print("Events = 0");                         // display Events = 0
        attachInterrupt(1, ISR, RISING);                 // configure interrupt INT1
}


void ISR()
{
        lcd.home();                                      // home the cursor
        cnt++;                                           // increment cnt
        lcd.print("Events = ");                          // display "Count = "
        lcd.print(cnt);                                  // display the Cnt in decimal
}


void loop()
{
}
```

Figure 7.73: Program Listing of the Project

and the development board. Variable **cnt** is declared and is set to 0. Inside the *setup()* function, the LCD is configured as 2 rows, 16 columns, the cursor is set to home position, and text **Events = 0** is displayed to start with. In addition, function **attachInterrupt** is used to configure external interrupt INT1 such that the program jumps to routine **ISR** when the state of the external interrupt pin INT1 (interrupt number 1) changes from 0 to 1, that is, when the button is pressed.

Figure 7.74 shows the project hardware setup.

7.17 Project 7.17 – Voltmeter

7.17.1 Project Description

In this project, the analogue-to-digital converter (ADC) of the PIC32MX320 microcontroller is used to design a voltmeter. The project measures the external analogue voltage in the range 0 to +3.3 V and displays it on the LCD.

Figure 7.74: Project Hardware Setup

The PIC32MX320 microcontroller has 10-bit multiplexed ADC with 16 channels. The chipKIT MX3 development board provides access to 11 of these ADCs through the Pmod connectors. Because the ADC is 10-bit wide, the converted voltage has digital value in the range 0–1023.

The analogue inputs are accessed using the built-in function **analogRead()**. The analogue input pin number is specified using the symbols A0–A10 (the digital pin numbers 0–10 can also be used if desired).

The Pmod connector references, logical I/O pin numbers, and microcontroller pin names of the analogue channels are given as follows:

Analogue Channel	Pmod Connector	Digital Pin Number	Microcontroller Pin
A0	JC-01	16	RB8
A1	JC-04	19	RB14
A2	JC-07	20	RB0
A3	JC-08	21	RB1
A4	JD-01	24	RB2
A5	JD-04	27	RB9
A6	JD-07	28	RB12
A7	JD-10	31	RB13
A8	JE-08	37	RB5
A9	JE-09	38	RB4
A10	JE-10	39	RB3

In this project, analogue channel A0 is used. The external analogue voltage is applied directly to this pin.

The lower reference voltage for the ADC is Vref−, while the upper reference voltage is Vref+. The reference voltage can be either external or internal. When internal reference is used, Vref− is connected to Vss and Vref+ is connected to VDD. Thus, the reference voltage is +3.3 V and an input voltage of +3.3 V will convert to digital value 1023. When external reference is used, an external voltage source can be connected to the Vref+ input. In this case, Vref− is shared with AN1 (RB1), and Vref+ is shared with AN0 (RB0).

In this project, the internal voltage reference is used. Built-in function **analogReference()** is used to select the ADC reference voltage. The following reference voltage values can be selected:

- DEFAULT (Vref− = 0 and Vref+ = +3.3 V)
- INTERNAL (same as DEFAULT)
- EXTERNAL (Vref− = 0 and Vref+ = voltage at pin A2)
- EXTMINUS (Vref− = voltage at pin A3 and Vref+ = +3.3 V)
- EXTPLUSMINUS (Vref− = voltage at pin A3 and Vref+ = voltage at pin A2)

Figure 7.75 shows the block diagram of the project.

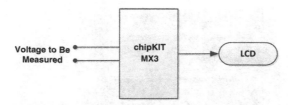

Figure 7.75: Block Diagram of the Project

7.17.2 Project Hardware

The voltage to be measured is applied to analogue input A0 (Pmod connector JC-01, logical I/O number 16, pin RB8). As shown in Figure 7.76, the PmodCLP LCD module is connected to Pmod connectors JA and JB as in the previous projects. External analogue voltages up to +3.3 V can be measured and displayed on the LCD. The breadboard Pmod module PmodBB is used to connect the voltage to be measured to Pmod connector JC-01. Pins JC-01 and JC-05 (ground) of connector JC are used to measure the external voltage. Figure 7.77 shows the hardware setup.

7.17.3 Project PDL

The operation of the project is described in the PDL shown in Figure 7.78. At the beginning of the program, the interface between the LCD and the chipKIT MX3 board is specified. A variable called **lcd** of type **LiquidCrystal** is declared. The conversion factor is then defined that converts the read analogue input voltage into millivolts. A heading is displayed, and the ADC reference voltage is set to INTERNAL. The program then reads the analogue voltage and displays on the LCD every second.

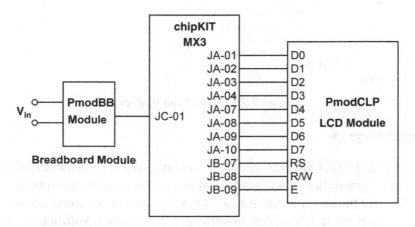

Figure 7.76: The Project Hardware

Figure 7.77: The Hardware Setup

BEGIN
  Define the interface between the LCD and the development board
  Define conversion factor
  Display heading
  Configure ADC reference voltage as INTERNAL
  DO FOREVER
    Read channel 0 voltage
    Convert to millivolts
    Display voltage
    Wait one second
  ENDDO
END

Figure 7.78: PDL of the Project

7.17.4 Project Program

The program is called VOLTMETER, and the program listing is shown in Figure 7.79. At the beginning of the program, function **LiquidCrystal** is called to define the interface between the LCD module and the development board. The conversion factor **conv**, when multiplied by the digital value, converts it into voltage in millivolts. The heading **Voltmeter...** is displayed on the top row of the LCD, and the ADC reference voltage is set to INTERNAL (+3.3 V).

```
/*==============================================================================
                               VOLTMETER
                               =========
```

In this project, an LCD is connected to the chipKIT MX3 development board as in the previous project. i.e. to Pmod connectors JA and JPB.

In addition, the PmodBB module is used to route the analog voltage to be measured to pin JC-01 (analog input A0) of the chipKIT MX3 development board.

The project measures the external voltage (0 to +3.3V) and displays it on the LCD.

```
      Author:        Dogan Ibrahim
      Date:          May, 2014
      File:          VOLTMETER
      Board:         chipKIT MX3
==============================================================================*/
//
// Include the LiquidCrystal library in the program
//
#include <LiquidCrystal.h>

constint Channel = 0;
const float conv = 3300 / 1024;
int DAC;
float mV;

//
// Initialize the LCD library with the numbers of the interface pins
//
  LiquidCrystal lcd(12, 13, 14, 0, 1, 2, 3, 4, 5, 6, 7);            // define LCD interface

void setup()
{
        lcd.begin(16, 2);                                          // 16 column,2 row
        lcd.home();                                                // home cursor
        lcd.print("Voltmeter...");                                 // display heading
        analogReference(INTERNAL);                                 // set reference voltage

}




void loop()
{
        DAC = analogRead(Channel);                                 // read channel 0 data
        lcd.setCursor(0, 1);                                       // col=0,row=1
        mV = DAC * conv;                                           // convert to mV
        lcd.print(mV);                                             // display mV
        lcd.print("mV");                                           // display "V"
        delay(1000);                                               // wait 1 second
        lcd.setCursor(0, 1);                                       // position to second row
        lcd.print("            ");                                 // clear second row
}
```

Figure 7.79: Program Listing of the Project

Inside the main program loop, the analogue voltage is read from channel 0 using function **analogRead(Channel)**, where **Channel** is set to 0. The voltage in millivolts is then found by multiplying with **conv** and is displayed on the second row of the LCD. This process is repeated forever after 1-s delay.

The design given in this circuit can be modified to measure higher input voltages by using potential divider resistors at the input circuit. For example, by attenuating the input voltage by about a factor of 10, input voltages up to 33 V can be measured. It is also recommended to use diodes at the input circuit to protect against excessive voltages and against reversed input polarity.

In Figure 7.80, the voltage at the ADC input is:

$$\mathrm{ADC} = V_{\mathrm{in}} \times \frac{10\mathrm{k}}{10\mathrm{k}+100\mathrm{k}} = 0.0909V_{\mathrm{in}}$$

Thus, the maximum input voltage that can be measured is 3.3 V/0.0909 ≈ 36 V.

7.18 Project 7.18 – Temperature Measurement

7.18.1 Project Description

In this project, the ADC of the PIC32MX320 microcontroller is used to design a digital thermometer. The project measures the ambient temperature using a LM35DZ-type temperature sensor chip and then displays the temperature on the LCD in the following format:

    Thermometer…
    nn.nnC

In this project, analogue channel A0 is used as in the previous project. The output pin of the temperature sensor chip is directly connected to this channel. In this project, the internal voltage reference is used. Built-in function **analogReference()** is used to select the ADC reference voltage as INTERNAL (i.e., +3.3 V).

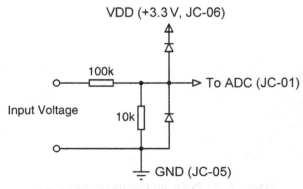

Figure 7.80: Increasing the Input Range

Figure 7.81: LM35DZ Temperature Sensor Pin Configuration

LM35DZ is a three-pin temperature sensor chip (see Figure 7.81 for pin configuration), having the following features:

- Operation in the range of 4–39 V
- Measurement in the range of $-55°C$ to $+150°C$
- Linear output voltage at 10 mV/°C
- Less than 60 μA current consumption
- Self-heating less than 0.1°C
- Plastic TO-92 transistor packaging

Figure 7.82 shows the block diagram of the project.

7.18.2 Project Hardware

The LM35DZ temperature sensor chip is connected to analogue input A0 (Pmod connector JC-01, logical I/O number 16, pin RB8). As shown in Figure 7.83, the PmodCLP LCD module is connected to Pmod connectors JA and JB as in the previous projects. The breadboard Pmod module PmodBB is used to connect the LM35DZ chip to Pmod connector JC-01. Pmod JPC connector power supply selection jumper is set to +5 V since the LM35DZ requires at least 4 V for its operation.

Figure 7.84 shows the hardware setup.

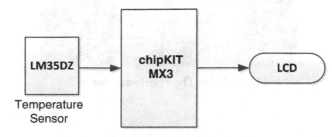

Figure 7.82: Block Diagram of the Project

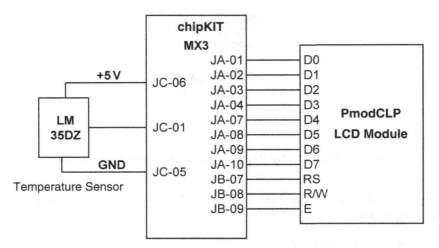

Figure 7.83: The Project Hardware

Figure 7.84: The Hardware Setup

```
BEGIN
        Define the interface between the LCD and the development board
        Define conversion factor
        Display heading
        Configure ADC reference voltage as INTERNAL
        DO FOREVER
                Read channel 0 voltage
                Convert to millivolts
                Divide by 10 to find the ambient temperature
                Display the temperature
                Wait one second
        ENDDO
END
```

Figure 7.85: PDL of the Project

7.18.3 Project PDL

The operation of the project is described in the PDL shown in Figure 7.85. At the beginning of the program, the interface between the LCD and the chipKIT MX3 board is specified. A variable called **lcd** of type **LiquidCrystal** is declared. The conversion factor is then defined that converts the read analogue input voltage into millivolts. A heading is displayed, and the ADC reference voltage is set to INTERNAL. The program then reads the analogue voltage, converts into millivolts, and divides by 10 to find and display the ambient temperature on the LCD every second.

7.18.4 Project Program

The program is called THERMO1, and the program listing is shown in Figure 7.86. At the beginning of the program, function **LiquidCrystal** is called to define the interface between the LCD module and the development board. The conversion factor **conv**, when multiplied by the digital value, converts it into voltage in millivolts. The heading **Thermometer...** is displayed on the top row of the LCD, and the ADC reference voltage is set to INTERNAL (+3.3 V). Inside the main program loop, the analogue voltage is read from channel 0 using function **analogRead**(Channel), where **Channel** is set to 0. The voltage in millivolts is then found by multiplying with **conv** and then dividing by 10 to find the temperature in degrees centigrade. The temperature is displayed on the LCD every second.

7.19 Project 7.19 – Temperature Measurement With Serial Monitor Display
7.19.1 Project Description

This project is very similar to the previous project where the ambient temperature is measured and displayed every second. In this project, the temperature is displayed on the MPIDE Serial Monitor.

```
/*==============================================================================
                              THERMOMETER
                              ===========

In this project, an LCD is connected to the chipKIT MX3 development board as in the previous
project. i.e. to Pmod connectors JA and JPB.

In addition, the PmodBB module is used to route the analog voltage to be measured to pin JC-01
(analog input A0) of the chipKIT MX3 development board and a LM35DZ type temperature sensor
is connected to this analog input.

The project measures the ambient temperature and displays it on the LCD in the following format:

                  Thermometer...
                  nn.nnC

Author:        Dogan Ibrahim
Date:          May, 2014
File:          THERMO1
Board:         chipKIT MX3
==============================================================================*/
//
// Include the LiquidCrystal library in the program
//
#include <LiquidCrystal.h>

constint Channel = 0;
const float conv = 3300 / 1024;
int DAC;
float mV;

//
// Initialize the LCD library with the numbers of the interface pins
//
  LiquidCrystal lcd(12, 13, 14, 0, 1, 2, 3, 4, 5, 6, 7);              // define LCD interface

void setup()
{
        lcd.begin(16, 2);                                // 16 column,2 row
        lcd.home();                                      // home cursor
        lcd.print("Thermometer...");                     // display heading
        analogReference(INTERNAL);                       // set reference voltage

}
void loop()
{
        DAC = analogRead(Channel);                       // read channel 0 data
        lcd.setCursor(0, 1);                             // col=0,row=1
        mV = DAC * conv;                                 // convert to mV
        mV = mV / 10.0;                                  // convert to Degrees C
        lcd.print(mV);                                   // display mV
        lcd.print("C");                                  // display "V"
        delay(1000);                                     // wait 1 second
        lcd.setCursor(0, 1);                             // position to second row
        lcd.print("           ");                        // clear second row
}
```

Figure 7.86: Program Listing of the Project

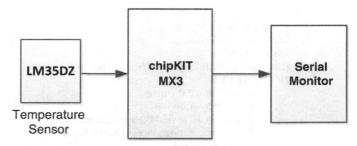

Figure 7.87: Block Diagram of the Project

In this project, the LM35DZ temperature sensor chip is used as in the previous project and is connected to analogue channel A0.

Figure 7.87 shows the block diagram of the project.

7.19.2 Project Hardware

As shown in Figure 7.88, the LM35DZ temperature sensor chip is connected to analogue input A0 (Pmod connector JC-01, logical I/O number 16, pin RB8). The breadboard Pmod module PmodBB is used to connect the LM35DZ chip to Pmod connector JC-01. Pmod JPC connector power supply selection jumper is set to +5 V since the LM35DZ requires at least 4 V for its operation.

7.19.3 Project PDL

The operation of the project is described in the PDL shown in Figure 7.89. At the beginning of the program, the Serial Monitor interface is initialised to 9600 Baud. A heading is displayed, and the ADC reference voltage is set to INTERNAL. The program then reads the

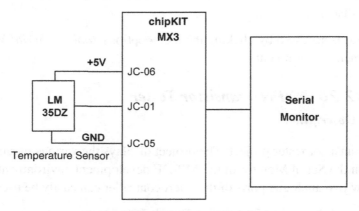

Figure 7.88: The Project Hardware

```
BEGIN
        Configure Serial Monitor
        Define conversion factor
        Display heading
        Configure ADC reference voltage as INTERNAL
        DO FOREVER
                Read channel 0 voltage
                Convert to millivolts
                Divide by 10 to find the ambient temperature
                Display the temperature on Serial Monitor
                Wait one second
        ENDDO
END
```

Figure 7.89: PDL of the Project

analogue voltage, converts into millivolts, and divides by 10 to find and display the ambient temperature on the Serial Monitor every second.

7.19.4 Project Program

The program is called THERMO2, and the program listing is shown in Figure 7.90. At the beginning of the program, the conversion factor **conv**, when multiplied by the digital value, converts it into voltage in millivolts. Serial Monitor is configured to operate at 9600 Baud using function **Serial.begin(9600)**. The heading **Thermometer...** is displayed on the top row of the Serial Monitor using function **Serial.print("Thermometer...\n")**, and the ADC reference voltage is set to INTERNAL (+3.3 V). Inside the main program loop, the analogue voltage is read from channel 0 using function **analogRead**(Channel), where **Channel** is set to 0. The voltage in millivolts is then found by multiplying with **conv** and then dividing by 10 to find the temperature in degrees centigrade. The temperature is displayed on the Serial Monitor every second as shown in Figure 7.91. Notice that a newline character (\n) is inserted at the end of the print command (the command **Serial.println()** could also be used to insert a newline after the display).

The Serial Monitor is accessed by clicking the menu options *Tools → Serial Monitor* on the MPIDE development environment.

7.20 Project 7.20 – NPN Transistor Tester
7.20.1 Project Description

This is an NPN transistor tester project. The project displays the various parameters of an NPN transistor on the Serial Monitor of the MPIDE development environment. The project demonstrates how two analogue ports of the microcontroller can easily be used.

The block diagram of the project is shown in Figure 7.92.

```
/*==============================================================================
                              THERMOMETER
                              ============

In this project the PmodBB module is used to route the analog voltage to be measured to pin JC-01
(analog input A0) of the chipKIT MX3 development board and a LM35DZ type temperature sensor
is connected to this analog input.

The project measures the ambient temperature and displays it on the Serial Monitor of the MPIDE
development environment every second in the following format:

        Thermometer...
        nn.nnC


Author:         Dogan Ibrahim
Date:           May, 2014
File:           THERMO2
Board:          chipKIT MX3
==============================================================================*/

constint Channel = 0;
const float conv = 3300 / 1024;
int DAC;
float mV;

void setup()
{
    Serial.begin(9600);                         // set serial speed
    Serial.print("Thermometer...\n");           // display heading and newline
    analogReference(INTERNAL);                  // set reference voltage

}




void loop()
{
        DAC = analogRead(Channel);              // read channel 0 data
        mV = DAC * conv;                        // convert to mV
        mV = mV / 10.0;                         // convert to Degrees C
        Serial.print(mV);                       // display mV on serial monitor
        Serial.print("C\n");                    // display "C" and newline
        delay(1000);                            // wait 1 second
}
```

Figure 7.90: Program Listing of the Project

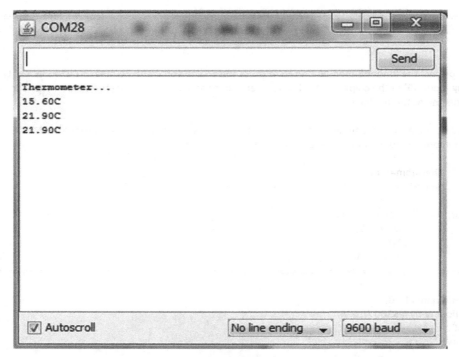

Figure 7.91: Displaying the Temperature on Serial Monitor

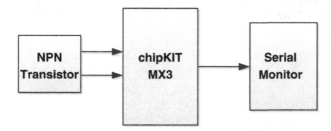

Figure 7.92: Block Diagram of the Project

7.20.2 Project Hardware

The transistor whose parameters are to be measured is connected to analogue ports A0 and A1 of the microcontroller as shown in the circuit diagram of Figure 7.93. The transistor is connected to the +3.3 V supply voltage (V_{CC}) through a 1k resistor. The base of the transistor is connected to the collector through a 100k resistor.

In reference to Figure 7.93 and assuming a silicon-type transistor, we can write the following equations about the various parameters of an NPN transistor:

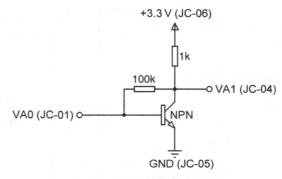

Figure 7.93: Project Circuit Diagram

$$V_{BE} = VA0$$

$$V_{CE} = VA1$$

$$I_C = \frac{V_{CC} - V_{CE}}{R_L}$$

$$I_B = \frac{V_{CE} - V_{BE}}{R_B}$$

$$\beta = \frac{I_C}{I_B}$$

$$I_E = I_B + I_C$$

In this project, R_L and R_B are chosen as 1k and 100k, respectively. The ADC reference voltage is chosen as +3.3 V (INTERNAL). All the currents are expressed in milliamperes, and all the voltages are expressed in volts.

As shown in Figure 7.94, the base of the transistor is connected to analogue input A0 (Pmod connector JC-01, logical I/O number 16, pin RB8) and the collector is connected to A1 (Pmod connector JC-04, logical I/O number 19, pin RB14). The breadboard Pmod module PmodBB is used to connect the transistor to be tested to Pmod connector JC. The hardware setup is shown in Figure 7.95.

7.20.3 Project PDL

The operation of the project is described in the PDL shown in Figure 7.96. At the beginning of the program, the Serial Monitor interface is initialised to 9600 Baud. A heading is displayed, and the ADC reference voltage is set to INTERNAL. The program then reads the base and collector analogue voltages, calculates the various currents and voltages in the circuit, and then displays them on the Serial Monitor.

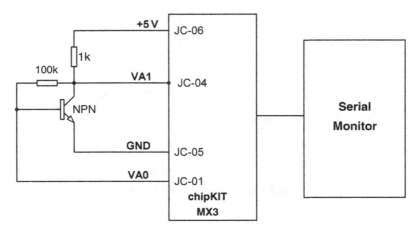

Figure 7.94: Project Hardware

Figure 7.95: Hardware Setup

BEGIN
    Configure Serial Monitor
    Define conversion factor and various constants
    Display heading
    Configure ADC reference voltage as INTERNAL
    DO FOREVER
        Read channel 0 (VA0) and channel 1 (VA1) voltages
        Calculate and display voltages and currents
        Wait here
    ENDDO
END

Figure 7.96: PDL of the Project

7.20.4 *Project Program*

The program is called NPN, and the program listing is shown in Figure 7.97. At the beginning of the program, various constants used in the program are declared. Then, the Serial Monitor is configured to operate at 9600 Baud. Inside the main program loop, all the required voltages and currents are calculated using the formulas given earlier and displayed via the Serial Monitor.

Figure 7.98 shows a typical output from the program (in this example, a BC108-type NPN transistor was used).

7.21 Project 7.21 – Writing to SD Card

7.21.1 *Project Description*

In this and the next few projects, we will be using SD cards as storage devices. But before going into the details of these projects, we should take a look at the basic principles and operation of SD card memory devices.

SD cards are commonly used in many electronic devices where large amount of nonvolatile data storage is required. Some application areas are:

- Digital cameras
- Camcorders
- Printers
- Laptop computers
- Global Positioning System (GPS) receivers
- Electronic games
- PDAs
- Mobile phones
- Embedded electronic systems

Figure 7.99 shows the picture of a typical SD card.

The SD card is a flash memory storage device designed to provide high-capacity, nonvolatile, and rewritable storage in small size. The memory capacities of the SD cards are increasing all the time. Currently, they are available in capacities from several gigabytes to over 128 GB. SD cards are in three sizes: *standard SD card*, *miniSD card*, and the *microSD card*. Table 7.6 lists the main specifications of different size cards.

SD card specifications are maintained by the *SD Card Association* that has over 600 members. MiniSD and microSD cards are electrically compatible with the standard SD cards, and they can be inserted in special adapters and used as standard SD cards in standard card slots.

```
/*=========================================================================
                            TRANSISTOR TESTER
                            =================

This is a transistor tester project. The project displays the basic parameters of an NPN type transistor.

In this project the PmodBB module is used so that connections can be made to ADC ports A0 and A1.
A0 is at Pmod connector JC-01 and A1 is at logical I/O port JC-04.

The transistor is supplied with +3.3V supply through the JPC connector. The base and the collector
Voltages are read through the ADC and the basic parameters are displayed on the Serial Monitor.

The following equations are used in the calculations (VA0 and VA1 are the measured base and collector
Voltages of the transistor):

VBE = VA0
VCE = VA1
IC = (VCC - VCE) / RL
IB = (VCE - VBE) / RB
Beta = IC / IB
IE = IB + IC

      Author:      Dogan Ibrahim
      Date:        May, 2014
      File:        NPN
      Board:       chipKIT  MX3
      ====================================================================*/
```

```
constint VA0 = 0;                                        // base ADC
constint VA1 = 1;                                        // collector ADC
constint RL = 1000;                                      // RB = 1K
const long RB = 100000;                                  // RB = 100K
                        // VCC = 3/3V
const float VCC = 3.3;
const float conv = 3.3 / 1024;
int VBE, VCE;
float VBEV, VCEV, IC, IB;

void setup()
{
    Serial.begin(9600);                                  // set serial speed
    Serial.print("NPN TRANSISTOR PARAMETERS\n");         // display heading and newline
    Serial.print("=========================\n");
    analogReference(INTERNAL);                           // set reference voltage
}

void loop()
{
        VBE = analogRead(VA0);                           // read channel 0 (VBE) volts
        VCE = analogRead(VA1);                           // read channel 1 (VCE) volts
        VBEV = VBE * conv;                               // convert VBE to volts
        VCEV = VCE * conv;                               // convert VCE to volts
        Serial.print("VBE = ");                          // display VBE =
        Serial.print(VBEV);                              // display VBE
        Serial.print("V\n");                             // display V
```

Figure 7.97: Program Listing *(Continued)*

```
        Serial.print("VCE = ");              // display VCE =
        Serial.print(VCEV);                  // display VCE
        Serial.print("V\n");                 // display V
        IC = (VCC-VCEV) / RL;                // calculate IC
        Serial.print("IC = ");               // display IC =
        Serial.print(IC*1000);               // display IC
        Serial.print("mA\n");                // display mA
        IB = (VCEV-VBEV) / RB;               // calculate IB
        Serial.print("IB = ");               // display IB =
        Serial.print(IB*1000);               // display IB
        Serial.print("mA\n");                // display mA
        Serial.print("Beta = ");             // display Beta =
        Serial.print(IC/IB);                 // display Beta
        Serial.print("\nIE = ");             // display IE =
        Serial.print((IC+IB)*1000);          // display IE
        Serial.print("mA\n");                // display mA
        while(1);                            // wait here
}
```

Figure 7.97: *(cont.)*

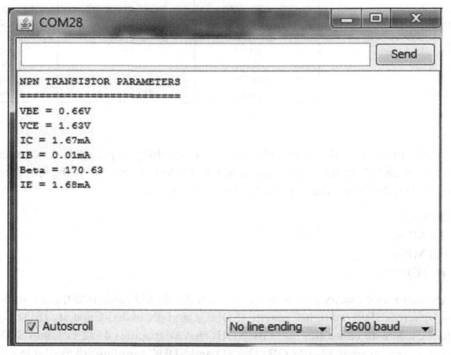

Figure 7.98: A Typical Output From the Program

Figure 7.99: A Typical SD Card

Table 7.6: Different size SD card specifications.

	Standard SD	miniSD	microSD
Dimensions (mm)	32 × 24 × 2.1	21.5 × 20 × 1.4	15 × 11 × 1
Card weight (g)	2.0	0.8	0.25
Operating voltage (V)	2.7–3.6	2.7–3.6	2.7–3.6
Write protect	Yes	No	No
Pins	9	11	8
Interface	SD or SPI	SD or SPI	SD or SPI
Current consumption (mA)	100 (write)	100 (write)	100 (write)

SD card speeds are measured in three different ways: in kilobytes per second, in megabytes per second, or in an "x" rating similar to that of CD-ROMS, where "x" is the speed corresponding to 150 KB/s. Thus, the various "x"-based speeds are:

- 4x: 600 KB/s
- 16x: 2.4 MB/s
- 40x: 6.0 MB/s
- 66x: 10 MB/s

As far as the memory capacity is concerned, we can divide SD cards into three families: Standard-Capacity (SDSC), High-Capacity (SDHC), and eXtended-Capacity (SDXC). SDSC are the older cards with capacities 1–2 GB. SDHC have capacities 4–32 GB, and SDXC cards have capacities greater than 32–128 GB. The SD and SDHC families are available in all three sizes, but the SDXC family is not available in the mini size.

In the projects in this book, we shall be using the standard SD cards only. The use of smaller-size SD cards is virtually the same and is not described here any further.

SD cards can be interfaced with microcontrollers using two different protocols: SD card protocol and the Serial Peripheral Interface (SPI) protocol. The SPI protocol is the most commonly used protocol and is the one used in the projects in this book. SPI bus is currently used by microcontroller interface circuits to talk to a variety of devices such as:

- Memory devices (SD cards)
- Sensors
- Real-time clocks
- Communications devices
- Displays

The advantages of the SPI bus are:

- Simple communication protocol
- Full duplex communication
- Very simple hardware interface

In addition, the disadvantages of the SPI bus are:

- Requires four pins
- No hardware flow control
- No slave acknowledgement

It is important to realise that there are no SPI standards governed by any international committee. As a result of this, there are several versions of the SPI bus implementation. In some applications, two data lines are combined into a single data line, thus reducing the line requirements into three. Some implementations have two clocks, one to capture (or display) data and another to clock it into the device. Also, in some implementations, the chip select line may be active-high rather than active-low.

The standard SD card has nine pins with the pin layout shown in Figure 7.100. Depending on the interface protocol used, pins have different functions. Table 7.7 gives the function of each pin both in SD mode and in SPI mode of operation.

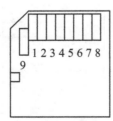

Figure 7.100: Standard SD Card Pin Layout

Table 7.7: Standard SD card pin definitions.

Pin	Name	SD Description	SPI Description
1	CD/DAT3/CS	Data line 3	Chip select
2	CMD/Datain	Command/response	Host to card command and data
3	VSS	Supply ground	Supply ground
4	VDD	Supply voltage	Supply voltage
5	CLK	Clock	Clock
6	VSS2	Supply voltage ground	Supply voltage ground
7	DAT0	Data line 0	Card to host data and status
8	DAT1	Data line 1	Reserved
9	DAT2	Data line 2	Reserved

Before going into details of the SD card projects, it is worthwhile to look at the operation of the SPI bus briefly.

Figure 7.101 shows a simplified block diagram of an SPI bus implementation with a master and three slave devices communicating over the SPI bus. In this figure, the slaves are selected by the slave select (SS) signals generated by the master. The clock and data lines are common to all the devices.

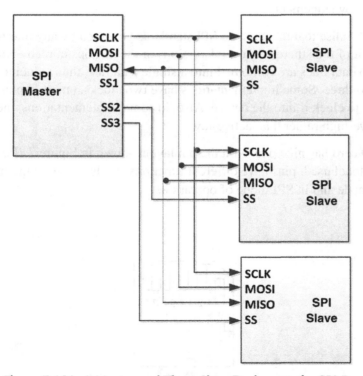

Figure 7.101: A Master and Three Slave Devices on the SPI Bus

7.21.2 chipKIT MX3 SPI Signals

The PIC32 microcontroller contains two SPI bus interfaces named SPI1 and SPI2. The SPI bus signals on the PIC32 are labelled as SS, serial data out (SDO), serial data in (SDI), and serial clock (SCL).

SPI1 is accessed via Pmod connector JB, and SPI2 is accessed via Pmod connector JE. The SPI bus interface pins are:

- JB-01: RD9 (used for SS output)
- JB-02: SDO1
- JB-03: SDI1
- JB-04: SCL1
- JE-01: SS2
- JE-02: MOSI
- JE-03: MISO
- JE-04: SCL2

SPI1 supports SPI master, but it can also be used as SPI slave by using external wiring, and when used as a slave, the SS1 signal is obtained from Pmod connector JD-01.

SPI2 can be used as an SPI master or slave. Jumpers JP6 and JP8 are used to select between master and slave by switching the microcontroller signals SDO2 and SDI2 between the SPI signals MOSI and MISO at the Pmod connector (M position for SPI master operation and S position for SPI slave operation).

The chipKIT SPI library or the Digilent DSPI library can be used to access the SPI bus when using the MPIDE development environment. The chipKIT library supports the SPI2 interface, while the DSPI library supports both SPI1 and SPI2 ports.

The SD card library (**sd.h**) distributed with the MPIDE development system uses the software-implemented SPI library with the following connections:

- JC-01: SS
- JC-02: SDO
- JC-03: SDI
- JC-04: SCL

7.21.3 Operation of the SD Card in SPI Mode

When the SD card is operated in SPI mode, only seven pins are used:

- Two power supply ground (pins 3 and 6)
- Power supply (pin 4)

- Chip select (pin 1)
- Data out (pin 7)
- Data in (pin 2)
- CLK (pin 5)

Three pins are used for the power supply, leaving four pins for the SPI mode of operation:

- Chip select (pin 1)
- Data out (pin 7)
- Data in (pin 2)
- CLK (pin 5)

At power-up, the SD card defaults to the SD bus protocol. The card is switched to the SPI mode if the CS signal is asserted during the reception of the reset command. When the card is in SPI mode, it only responds to SPI commands. The host may reset a card by switching the power supply off and on again.

Most high-level language compilers normally provide a library of commands for initialising, reading, and writing to SD cards. In general, it is not necessary to know the internal structure of an SD card before it can be used since the available library functions can easily be used. It is, however, important to have some knowledge about the internal structure of an SD card so that it can be used efficiently. In this section, we shall be looking briefly at the internal architecture and the operation of SD cards.

An SD card has a set of registers that provide information about the status of the card. When the card is operated in SPI mode, these registers are:

- Card Identification Register (CID)
- Card-Specific Data Register (CSD)
- SD Configuration Register (SCR)
- Operation Control Register (OCR)

The CID register consists of 16 bytes, and it contains the manufacturer ID, product name, product revision, card serial number, manufacturer date code, and a checksum byte.

The CSD register consists of 16 bytes, and it contains card-specific data such as the card data transfer rate, R/W block lengths, R/W currents, erase sector size, file format, write protection flags, checksum, etc.

The SCR register is 8 bytes long, and it contains information about the SD card's special features and capabilities such as the security support, data bus widths supported, etc.

The OCR register is only 4 bytes long, and it stores the VDD voltage profile of the card. The OCR shows the voltage range in which the card data can be accessed.

All SD card SPI commands are 6 bytes long with the MSB transmitted first. First byte is known as the "command" byte, and the remaining 5 bytes are "command arguments." Bit 6 of the command byte is set to "1," and the MSB bit is always "0." With the remaining 6 bits, we have 64 possible commands, named CMD0–CMD63. Some of the important commands are:

- CMD0: GO_IDLE_STATE (resets the SD card)
- CMD1: SEND_OP_COND (initialises the card)
- CMD9: SEND_CSD (gets CSD register data)
- CMD10: SEND_CID (gets CID register data)
- CMD16: SET_BLOCKLEN (selects a block length in bytes)
- CMD17: READ_SINGLE_BLOCK (reads a block of data)
- CMD24: WRITE_BLOCK (writes a block of data)
- CMD32: ERASE_WR_BLK_START_ADDR (sets the address of the first write block to be erased)
- CMD33: ERASE_WR_BLK_END_ADDR (sets the address of the last write block to be erased)
- CMD38: ERASE (erases all previously selected blocks)

In response to a command, the card sends a status byte known as R1. The MSB bit of this byte is always "0," and the other bits indicate various error conditions.

Reading data

SD card in SPI mode supports single block and multiple block read operations. The host should set the block length, and after a valid read command the card responds with a response token, followed by a data block, and a cyclic redundancy check (CRC). The block length can be between 1 and 512 bytes. The starting address can be any valid address range of the card.

In multiple block read operations, the card sends data blocks with each block having its own CRC attached to the end of the block.

Writing data

SD card in SPI mode supports single or multiple block write operations. After receiving a valid write command from the host, the card will respond with a response token, and will wait to receive a data block. A 1-byte "start block" token is added to the beginning of every data block. After receiving the data block, the card responds with a "data response" token and the card will be programmed as long as the data block has been received with no errors.

In multiple write operations, the host sends the data blocks one after the other, each preceded with a "start block" token. The card sends a response byte after receiving each data block.

A card can be inserted and removed from the bus without any damage. This is because all data transfer operations are protected by CRC codes and any bit changes as a result of inserting or removing a card can easily be detected. SD cards operate with a typical supply voltage of 2.7 V. The maximum allowed power supply voltage is 3.6 V. If the card is to be operated from a standard 5.0 V supply, a voltage regulator should be used to drop the voltage to 2.7 V.

Using an SD card requires the card to be inserted into a special card holder with external contacts (see Figure 7.102). Connections can then be made easily to the required card pins.

In this project, a file called **MYFILE.TXT** is created on the SD card and the following text is written to the file:

> **This book is about the chipKIT MX3 development board.**
> **The book gives the features of this development board.**
> **In addition, many projects are given in the book.**

The block diagram of the project is shown in Figure 7.103.

Figure 7.102: SD Card Holder

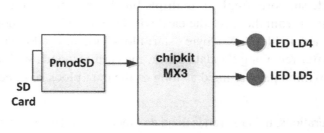

Figure 7.103: Block Diagram of the Project

7.21.4 Project Hardware

The Pmod module PmodSD is used in this project. As shown in Figure 7.104, PmodSD is a 2 × 6 pin module with a standard size SD card holder. The connection diagram of this module is shown in Figure 7.105.

When used in SPI data mode, only the pins at the top row (1–6) are used. The WP signal (pin 10) is the optional write protect and can be set by a switch on the SD card to prevent writing or erasing the card accidentally. The CD signal (pin 9) is optional and can be used by the microcontroller to indicate that a card is in the SD card slot.

In this project, the PmodSD module is connected to Pmod connector JC (see Figure 7.106). Therefore, the interface between the PmodSD module and the development board is as follows (notice that JC-01 is at logical I/O port number 16):

Pmod JPC Connector	SPI Signal Name	SD Card SPI Signal
JC-01	SS	CS
JC-02	SDO	DI
JC-03	SDI	DO
JC-04	SCL	CLK

Figure 7.107 shows the circuit diagram of the project (notice that LEDs LD4 and LD5 are not shown in this figure as they are mounted on the chipKIT MX3 board).

Figure 7.104: PmodSD Module

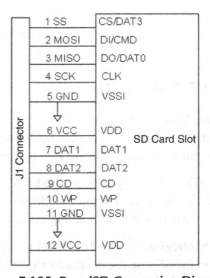

Figure 7.105: PmodSD Connection Diagram

Figure 7.106: Hardware Setup

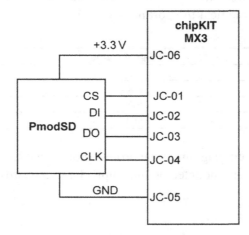

Figure 7.107: Project Circuit Diagram

7.21.5 Project PDL

Before looking at the programming details, it is worthwhile to see what type of SD card library functions are available for managing SD card–based applications. A brief summary of the commonly used SD card library functions is given below.

The SD card library supports both FAT16 and FAT32 file systems on standard SD and SDHC cards. Files can have names made up of up to 8 characters, followed by an extension made up of 3 characters (8.3 file naming convention).

SD.begin(cs): This function initialises the SD card library. The cs is optional and defaults to the SS line for the hardware SPI bus. The function returns True on success and False on failure.

SD.exists(filename): This function checks to see whether or not the specified file exists in the current directory. The forward slash character "/" can be used to look for the file in a directory. The function returns a True if file exists, or False if the file does not exist.

SD.mkdir(name): This function creates a folder on the SD card with the specified name. A subdirectory can be specified using forward slashes. A True is returned if the requested directory is created; otherwise, a False is returned.

SD.remove(filename): This function deletes the specified file on the SD card. A True is returned if the file is removed successfully; otherwise, a False is returned.

SD.rmdir(name): This function deletes a directory on the SD card. The directory must be empty before it can be removed. A True is returned if the operation is successful; otherwise, a False is returned.

SD.open(filename, mode): This function opens an existing file on the SD card. If the file does not exist, then a new file with the specified name will be created. Notice that the filename can contain directory names if required. The directory names must be specified using the forward slash character. Mode specifies whether the file will be opened for reading or writing. FILE_READ opens the file for reading, and the file pointer is positioned at the beginning of the file. FILE_WRITE opens the file for reading and writing. If the file exists, then the file pointer is positioned at the end of the file.

file.close(): This function writes any remaining data to the file and closes the file. *file* is an instance of the file class returned by **SD.open()**.

file.read(): This function is used to read a byte (next byte) from the file. *file* is an instance of the file class returned by **SD.open()**.

file.available(): This function checks if there are any bytes available for reading from the file. The number of bytes available for reading is returned as an integer. *file* is an instance of the file class returned by **SD.open()**.

file.write(data): This function is used to write a byte to the already opened file. *file* is an instance of the file class returned by **SD.open()**.

file.write(buffer, len): This function is used to write an array of characters of length **len** to the file. *file* is an instance of the file class returned by **SD.open()**. The number of bytes written to the file is returned as a byte.

file.flush(): This function ensures that all the bytes belonging to the file are physically written to the file (the **close** function automatically flushed the data before closing the file).

file.position(): This function returns (as an unsigned long) the position of the next location in the file where data is to be written to or read from. *file* is an instance of the file class returned by **SD.open()**.

file.size(): This function returns the size of the file in bytes. *file* is an instance of the file class returned by **SD.open()**.

file.print(data): This function writes data to the file. The data can be byte, integer, long, or string. *file* is an instance of the file class returned by **SD.open()**. An optional BASE parameter can be used to specify the base of the number to be printed. Valid BASE param-

```
        BEGIN
                Configure SD card Chip Select pin
                Configure LEDs LD4 and LD5 as outputs
                Turn OFF LD4 and LD5
                Start the SD card library
                IF error THEN
                        Turn ON LD4
                        EXIT
                ENDIF
                Open file MYFILE.TXT on the SD card
                IF successful THEN
                        Write the following text to the file
                        This book is about the chipKIT MX3 development board.
                        The book gives the features of this development board.
                        In addition, many projects are given in the book.
                ELSE
                        Turn ON LD4 and LD5
                ENDIF

                DO FOREVER

                ENDDO
        END
```

Figure 7.108: PDL of the Project

eters are BIN (binary), DEC (decimal), OCT (octal), or HEX (hexadecimal). The number of bytes written to the file is returned by the function.

file.println(data): This function is similar to **file.print(data)**, but here newline characters (carriage return and line feed) are also written to the file.

Figure 7.108 shows the PDL of the project. At the beginning, the chip select is configured as logical I/O port number 16 (JC-01). Then the two on-board LEDs LD4 and LD5 are configured as outputs. These LEDs are used to indicate error conditions. The SD card library is then started. If there is an error in starting the library, then LD4 is turned ON and the program stops. Otherwise, file **MYFILE.TXT** is created on the SD card and the following text is written inside this file. If the file cannot be opened, then both LD4 and LD5 are turned ON to indicate an error condition:

This book is about the chipKIT MX3 development board.
The book gives the features of this development board.
In addition, many projects are given in the book.

7.21.6 Project Program

The header file **sd.h** must be included at the beginning of a program that uses the SD card library. The program is called FILEW, and the program listing is shown in Figure 7.109. At the beginning of the program, the chip select pin (logical I/O port number 16) and the LD4 (logical I/O port number 42) and LD5 (logical I/O port number 43) are configured as outputs. The remainder of the program is executed inside the *setup()* routine. Here, both LEDs are turned OFF initially to indicate that there are no errors to start with. Then, the SD card library is initialised by calling function **SD.begin**(). If there is an error in initialising this library, then LD4 is turned ON and the program stops. Otherwise, file **MYFILE.TXT** is created on the SD card and the required text is written inside this file. If the file cannot be opened, then both LEDs LD4 and LD5 are turned ON to indicate the error condition.

```
/*===============================================================================
                              WRITING TO SD CARD
                              ==================

This project is about writing text to an SD card. The SD card module PmodSDis plugged-in to
connector JPC of the chipKIT MX3 development board.

The project creates a file called MYFILE.TXT and writes the following text to this file:

        This book is about the chipKIT MX3 development board.
        The book gives the features of this development board.
        In addition, many projects are given in the book.

Author:     Dogan Ibrahim
Date:       May, 2014
File:       FILEW
Board:      chipKIT MX3
=============================================================================*/

#include <SD.h>                                 // include SD card library

File myFile;
const int chipSelect_SD = 16;                   // CS is at logical port 16
const int LD4 = 42;                             // LD4 is at logical port 42
const int LD5 = 43;                             // LD5 is at logical port 43

void setup()
{
        pinMode(LD4, OUTPUT);                   // set LD4 as output
        pinMode(LD5, OUTPUT);                   // set LD5 as output
        digitalWrite(LD4, LOW);                 // turn OFF LD4 to start with
        digitalWrite(LD5, LOW);                 // turn OFF LD5 to start with

        pinMode(chipSelect_SD, OUTPUT);         // set CS as output
        digitalWrite(chipSelect_SD, HIGH);

        if(!SD.begin(chipSelect_SD))            // start the library
        {
                digitalWrite(LD4, HIGH);        // Turn ON LD4 if error
                return;
        }
```

Figure 7.109: Program Listing *(Continued)*

```
//
// Create file MYFILE.TXT on the SD card
//
        myFile = SD.open("MYFILE.TXT", FILE_WRITE);        // open file MYFILE.TXT
//
// If the file create successfully then write the text to the file
//
        if (myFile)                                        // if opened successfully
        {
            myFile.println("This book is about the chipKIT MX3 development board.");
            myFile.println("The book gives the features of this development board.");
            myFile.println("In addition, many projects are given in the book.");
            myFile.close();
        }
        else
        {
            digitalWrite(LD4, HIGH);                        // turn ON LD4
            digitalWrite(LD5, HIGH);                        // turn ON LD5
        }
}

void loop()
{
}
```

Figure 7.109: *(cont.)*

Figure 7.110 shows contents of file MYFILE.TXT displayed on the PC using the Notepad program after inserting the card to the PC via an SD card reader.

7.22 Project 7.22 – Reading From SD Card and Displaying on Serial Monitor

7.22.1 Project Description

This project is similar to the previous one, but here the file MYFILE.TXT created in the previous project is opened for reading and its contents are displayed on the Serial Monitor.

The block diagram of the project is shown in Figure 7.111.

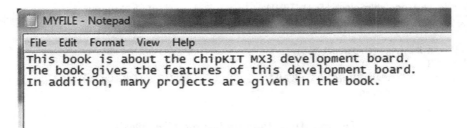

Figure 7.110: Contents of File MYFILE.TXT

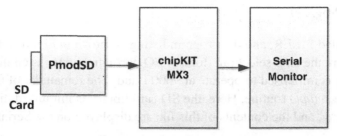

Figure 7.111: Block Diagram of the Project

7.22.2 Project Hardware

As in the previous project, the PmodSD module is connected to Pmod connector JPC (see Figures 7.106 and 7.107).

7.22.3 Project PDL

The project PDL is shown in Figure 7.112. At the beginning, the chip select is configured as logical I/O port number 16 (JC-01), and Serial Monitor is also initialised. The SD card library is then started. If there is an error in starting the library, then an appropriate message is sent to the Serial Monitor and the program stops. Otherwise, file **MYFILE.TXT** is opened for reading and its contents are read and displayed on the Serial Monitor. If the file cannot be opened, then an appropriate message is sent to the Serial Monitor.

```
BEGIN
        Configure SD card Chip Select pin
        Configure Serial Monitor
        Start the SD card library
        IF error THEN
                Send an error message to Serial Monitor
                EXIT
        ENDIF
        Open file MYFILE.TXT on the SD card
        IF successful THEN
                Read contents of the file
                Display contents of the file on Serial Monitor
        ELSE
                Send an error message to Serial Monitor
        ENDIF

        DO FOREVER

        ENDDO
END
```

Figure 7.112: PDL of the Project

7.22.4 Project Program

The program is called FILER, and the program listing is shown in Figure 7.113. At the beginning of the program, the chip select pin (logical I/O port number 16) is configured as output, and Serial Monitor is initialised to operate at 9600 Baud. The remainder of the program is executed inside the *setup()* routine. Here, the SD card library is initialised, file MYFILE.TXT is opened for reading, and the contents of this file are displayed on the Serial Monitor.

Figure 7.114 shows the Serial Monitor displaying contents of the file.

7.23 Project 7.23 – Temperature Data Logging on SD Card

7.23.1 Project Description

This project measures the ambient temperature as in Project 7.19. In this project, the temperature is read every 5 s and is stored on an SD card in a file called **TEMPS.TXT**. The program runs for 1 min where 12 records are written to the file.

The block diagram of the project is shown in Figure 7.115. As in Project 7.19, the temperature sensor chip LM35DZ is used in this project. LM35DZ is a three-pin temperature sensor chip whose output voltage is proportional to the ambient temperature and is given by: $V_o = 10$ mV/°C. Thus, for example, at 20°C the output voltage is 200 mV, at 30°C the output voltage is 300 mV, and so on.

7.23.2 Project Hardware

As in Project 7.22, the PmodSD module is connected to Pmod connector JC (see Figures 7.106 and 7.107). When using the MPIDE development environment with the PIC32 microcontrollers, the available analogue input channels are given as in the following table:

Analogue Channel	Pmod Connector	Digital Pin Number	Microcontroller Pin
A0	JC-01	16	RB8
A1	JC-04	19	RB14
A2	JC-07	20	RB0
A3	JC-08	21	RB1
A4	JD-01	24	RB2
A5	JD-04	27	RB9
A6	JD-07	28	RB12
A7	JD-10	31	RB13
A8	JE-08	37	RB5
A9	JE-09	38	RB4
A10	JE-10	39	RB3

```
/*==============================================================================
                        READING FROM SD CARD
                        ====================
This project is about reading text from an SD card and then displaying the text on the Serial Monitor.
The SD card module PmodSD is plugged-in to connector JPC.

The project opens file MYFILE.TXT on the SDcard and reads its contents.

Author:         Dogan Ibrahim
Date:           May, 2014
File:           FILER
Board:          chipKIT MX3
==============================================================================*/
#include <SD.h>                                          // include SD card library

File myFile;
const int chipSelect_SD = 16;                            // CS is at logical port 16

void setup()
{
        Serial.begin(9600);                              // configure Serial Monitor
        pinMode(chipSelect_SD, OUTPUT);                  // set CS as output
        digitalWrite(chipSelect_SD, HIGH);

        if(!SD.begin(chipSelect_SD))                     // start the library
        {
                Serial.println("Error initializing SD card library...");
                return;
        }
//
// Create file MYFILE.TXT on the SD card
//
        myFile = SD.open("MYFILE.TXT", FILE_READ);       // open file MYFILE.TXT
//
// If the file create successfully then write the text to the file
//
        if (myFile)                                      // if opened successfully
        {
                while(myFile.available())                // if any bytes...
                {
                        Serial.write(myFile.read());     // read and display on Serial Monitor
                }
                myFile.close();
        }
        else
        {
                Serial.println("Error opening file...");  // display error message
        }
}


void loop()
{
}
```

Figure 7.113: Program Listing

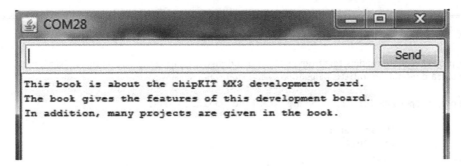

Figure 7.114: The Serial Monitor

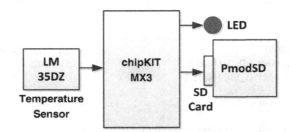

Figure 7.115: Block Diagram of the Project

Because Pmod connector JC is used for the SD card, the output of the temperature sensor chip is connected to analogue port A4 (Pmod connector JD-01). The Pmod breadboard module PmodBB is used to interface to the LM35DZ temperature sensor chip.

Figure 7.116 shows the project hardware setup. The circuit diagram of the project is shown in Figure 7.117 (LED LD4 is not shown here since it is on the chipKIT MX4 board).

7.23.3 Project PDL

The project PDL is shown in Figure 7.118. At the beginning, the chip select is configured as logical I/O port number 16 (JC-01) and the analogue-to-digital conversion factor is declared. The SD card library is then started. If there is an error in starting the library, then LED LD4 is turned ON and the program stops. Otherwise, file **TEMPS.TXT** is created and the temperature is written to the file every 5 s. The data is recorded for up to 1 min (i.e., 12 records), and after this time the file is closed and the program stops. If the file cannot be created, then the LED LD4 is turned ON to indicate the error condition.

Figure 7.116: Project Hardware Setup

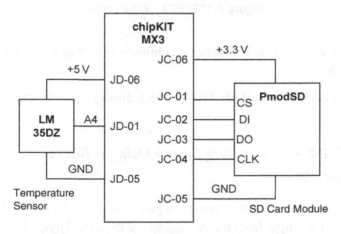

Figure 7.117: Circuit Diagram of the Project

7.23.4 Project Program

The program is called LOGGER, and the program listing is shown in Figure 7.119. At the beginning of the program, the chip select pin (logical I/O port number 16) is configured as an output. The SD card library is started using SD library function *SD.begin()*. If the library is started successfully, then file TEMPS.TXT is created on the SD card and the heading **TEMPERATURE DATA LOGGER** is written to the file. The program then reads the temperature

```
BEGIN
        Configure SD card Chip Select pin
        Configure ADC
        Start the SD card library
        IF error THEN
                Turn ON LD4
                EXIT
        ENDIF
        Open file TEMPS.TXT on the SD card
        IF successful THEN
                Write heading to the file
        ELSE
                Turn ON LD4
        ENDIF

        DO 12 TIMES
                Read analog channel 4
                Convert into millivolts
                Convert into temperature in degrees Centigrade
                Write to file
                Wait 5 seconds
        ENDDO
END
```

Figure 7.118: PDL of the Project

every 5 s and writes it to the file. This process continues for 1 min where a total of 12 records are written to the file.

Figure 7.120 shows the contents of file TEMPS.TXT displayed on the PC using the Notepad program.

7.24 Project 7.24 – Generating Sound Using a Buzzer

7.24.1 Project Description

This project uses a small piezoelectric buzzer to generate sound. In this project, the SOS Morse code (...---...) is output from the buzzer, where the sound frequency is set to 440 Hz. The basic dot duration is set to 150 ms.

The block diagram of the project is shown in Figure 7.121.

The Morse code dot and dash timings are as follows:

- Basic unit is the dot time.
- Dash time is three times the dot time.
- The gap between dots and dashes is equal to one dot time.
- The gap between letters is equal to three dot times.
- The gap between words is seven dot times.

```
/*===============================================================================
                         TEMPERATURE DATA LOGGER
                         ========================
```

In this project the PmodSD module is used to interface an SD card to the chipKIT MX3 development
board. In addition, PmodBB module is used to route the analogvoltage to be measured to pin JD-01
(analog input A4) of the board and a LM35DZ type temperature sensoris connected to this analog input.

The project creates a file called TEMPS.TXT on the SD card. It then measures the ambient temperature
every 5 seconds and writes the data to the file. The program continues for one minute where 12
records are written to the file. The file is closed at the end.

Author: Dogan Ibrahim
Date: May, 2014
File: LOGGER
Board: chipKIT MX3

```
===============================================================================*/
#include <SD.h>                                          // include SD card library
//
// SD card constants and variables
//
File myFile;
constintchipSelect_SD = 16;                              // CS is at logical port 16
constint LD4 = 42;                                       // LD4 is at logical port 42
//
// Temperature sensor constants and variables
//
constint Channel = 4;                                    // LM35DZ on analog A4
const float conv = 3300 / 1024;                          // ADC conversion factor
int DAC;
float mV;
unsigned char k;

void setup()
{
        pinMode(LD4, OUTPUT);                            // set LD4 as output
        digitalWrite(LD4, LOW);                          // turn OFF LD4 to start with
        pinMode(chipSelect_SD, OUTPUT);                  // set CS as output
        digitalWrite(chipSelect_SD, HIGH);
        analogReference(INTERNAL);                       // set reference voltage

        if(!SD.begin(chipSelect_SD))                     // start the SD library
        {                                                // if error...
                digitalWrite(LD4, HIGH);                 // Turn ON LD4 if error
                return;
        }
```

Figure 7.119: Program Listing *(Continued)*

```
//
// Create file TEMPS.TXT on the SD card
//
        myFile = SD.open("TEMPS.TXT", FILE_WRITE);         // open file TEMPS.TXT
        if(!myFile)                                         // if error...
        {
                digitalWrite(LD4, HIGH);                    // turn LD4 ON
                return;
        }
        myFile.println("TEMPERATURE DATA LOGGER");          // print heading
        myFile.println("=========================");
        myFile.println("");
}


void loop()
{
        for(k = 0; k < 12; k++)                             // do 12 times
        {
                DAC = analogRead(Channel);                  // read channel 4 data
                mV = DAC * conv;                            // convert to mV
                mV = mV / 10.0;                             // convert to Degrees C
                myFile.print(mV);                           // write to file
                myFile.println("C");                        // write "C" to file
                delay(5000);                                // wait 5 seconds
        }
        myFile.close();                                     // close the file
        while(1);                                           // wait here forever
}
```

Figure 7.119: *(cont.)*

7.24.2 Project Hardware

As shown in the circuit diagram in Figure 7.122, a small piezoelectric buzzer is connected to logical port pin 0 (Pmod connector JA-01) through a switching transistor. The buzzer sounds when the transistor is turned ON.

7.24.3 Project PDL

The MPIDE built-in sound functions *tone()* and *noTone()* are used in this project. As mentioned earlier, the *tone(pin, frequency, duration)* function has three arguments. The first argument is the pin number used, second argument is the frequency of the tone to be generated (in Hz), and duration is the tone duration (in milliseconds). The duration is optional and can be omitted if desired. Function *noTone()* has no arguments, and calling this function stops generating tone.

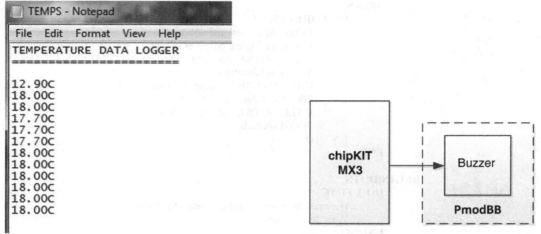

Figure 7.120: Contents of File TEMPS.TXT

Figure 7.121: Block Diagram of the Project

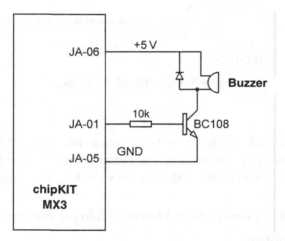

Figure 7.122: Circuit Diagram of the Project

The PDL of the project is shown in Figure 7.123. In this project, the piezoelectric buzzer is connected to logical port number 0 (Pmod connector JA-01) using the PmodBB breadboard module. The program configures the port as output and then calls to functions called DOTS and DASHES to generate the required SOS sound on the buzzer.

7.24.4 Project Program

The program is called SOS, and the program listing is shown in Figure 7.124. At the beginning of the program, logical port 0 is configured as output. Inside the main program, functions

```
BEGIN
    DO FOREVER
                Define frequency and dot duration
                Configure logical port 0 as output
                CALL DOTS to generate 3 dots
                Wait 3 dot duration
                CALL DASHES to generate 3 dashes
                Wait 3 dot duration
                CALL DOTS to generate 3 dots
                Wait 4 seconds
    ENNDO
END

BEGIN/DOTS
    DO 3 TIMES
                Generate tone with duration equal to dot time
                Wait one dot time
    ENDDO
END/DOTS

BEGIN/DASHES
    DO 3 TIMES
                Generate tone with duration equal to dash time
                Wait one dot time
    ENDDO
END/DASHES
```

Figure 7.123: PDL of the Project

DOTS and DASHES are called to generate the three dots and three dashes for the SOS signal. Function DOTS generates three short tones with durations equal to the basic dot time. Similarly, function DASHES generates three short notes with durations equal to three dot times.

7.25 Project 7.25 – Generating Melody Using a Buzzer

7.25.1 Project Description

This project shows how sound with different frequencies can be generated using a simple buzzer. The project shows how the simple melody **Happy Birthday** can be played using a buzzer.

The block diagram of the project is same as in Figure 7.121.

7.25.2 Project Hardware

The project hardware is same as in Figure 7.122.

```
/*===============================================================================
                        GENERATING SOUND ON A BUZZER
                        ============================
```

In this project a small piezoelectric buzzer is connected to logical port 0 (Pmod connector JA-01) through a transistor switch. The buzzer generates tone when a logical 1 is sent to the transistor. The SOS Morse code sound (...---...) is generated in this project.

Function DOTS generates 3 dot sounds, and function DASHES generates 3 dash sounds. The SOS code is sent with 4 second delay between each output.

The transistor and the buzzer are connected to the chipKIT MX3 development board using the PmodBB breadboard module.

```
Author:          Dogan Ibrahim
Date:            May, 2014
File:            SOS
Board:           chipKIT MX3
===============================================================================*/
int buzzer = 0;                                    // buzzer port
int frequency = 440;                               // note frequency
int dot_duration = 150;                            // dot duration is 150ms

void setup()
{
        pinMode(buzzer, OUTPUT);                   // set buzzer as output
        digitalWrite(buzzer, LOW);                 // turn OFF buzzer to start with
}


//
// This function generates 3 dot sounds
//
void DOTS()
{
        for(int k = 0; k < 3; k++)                 // 3 outputs
        {
                tone(buzzer, frequency);
                delay(dot_duration);               // wait dot duration
                noTone(buzzer);                    // stop tone
                delay(dot_duration);               // delay dot time
        }
}
```

Figure 7.124: Program Listing *(Continued)*

```
//
// This function generates 3 dash sounds
//
void DASHES()
{
        for(intk = 0; k < 3; k++)                        // 3 outputs
        {
                tone(buzzer, frequency);
                delay(3*dot_duration);                   // wait 3*dot duration
                noTone(buzzer);                          // stop tone
                delay(dot_duration);                     // delay dot time
        }
}


void loop()
{
        DOTS();                                          // generate 3 dot tones
        delay(3*dot_duration);                           // wait interval
        DASHES();                                        // generate 3 dash tones
        delay(3*dot_duration);                           // delay 3*dot time
        DOTS();                                          // generates 3 dot tones
        delay(4000);                                     // wait 4 second
}
```

Figure 7.124: *(cont.)*

7.25.3 Project PDL

When playing a melody, each note is played for a certain duration and with a certain frequency. In addition, a certain gap is necessary between two successive notes. The frequencies of the musical notes starting from middle C (i.e., C4) are given below. The harmonic of a note is obtained by doubling the frequency. For example, the frequency of C5 is $2 \times 262 = 524$ Hz.

Notes	C4	C4#	D4	D4#	E4	F4	F4#	G4	G4#	A4	A4#	B4
Frequency (Hz)	261.63	277.18	293.66	311.13	329.63	349.23	370	392	415.3	440	466.16	493.88

In order to play the tune of a song, we need to know its musical notes. Each note is played for certain duration, and there is a certain time gap between two successive notes.

In this project, we will be generating the classic **Happy Birthday** melody and thus we need to know the notes and their durations. These are given in the following table where the durations are in milliseconds and should be multiplied by 400 to give correct values:

Note	C4	C4	D4	C4	F4	E4	C4	C4	D4	C4	G4	F4	C4	C4	C5	A4	F4	E4	D4	A4#	A4#	A4	F4	G4	F4
Duration	1	1	2	2	2	3	1	1	2	2	2	3	1	1	2	2	2	2	2	1	1	2	2	2	4

```
BEGIN
        Create Notes and Durations tables
        Configure logical port 0 for output
        DO FOREVER
            DO for all notes
                    Play the note with specified duration
                    Delay 100ms
            ENDDO
        ENDDO
        Wait 3 seconds
END
```

Figure 7.125: PDL of the Project

The PDL of the project is shown in Figure 7.125. Basically two tables are used to store the notes and their corresponding durations. Then the *tone( )* function is called in a loop to play all the notes. The melody repeats after 3-s delay.

7.25.4 Project Program

The program is called MELODY, and its listing is given in Figure 7.126. At the beginning of the program, logical port 0 is configured for output and tables *Notes* and *Durations* store the frequencies (Hz) and durations (1/400 ms) of each note, respectively. Inside the main program loop, each note is played with the specified duration. A 100-ms gap is used between each note. The durations and the gap can be changed to increase or decrease the speed of the melody.

7.26 Project 7.26 – Using an Audio Amplifier

7.26.1 Project Description

This project is similar to Project 7.25, but here an audio amplifier is used to amplify the tones and output on a speaker.

Buzzers are normally used as alarm warning devices, and they do not produce tones at different frequencies. In this project, an audio amplifier is used to generate good-quality tones for our melody.

The block diagram of the project is shown in Figure 7.127.

7.26.2 Project Hardware

In this project, the audio amplifier module called AUDIO AMP manufactured by mikroElektronika (www.mikroe.com) is used. As shown in Figure 7.128, this is a small audio amplifier board that operates with +5 V supply. Audio signal is fed from one end of the board through a

```
/*================================================================================
                    GENERATING MELODY ON A BUZZER
                    ==============================

In this project a small piezoelectric buzzer is connected to logical port 0 (Pmod connector JA-01)
through a transistor switch. The buzzer operates when a logic 1 signal is sent to it.

The buzzer generates the popular melody HAPPY BIRTHDAY.

The transistor and the buzzer are connected to the chipKIT MX3 development board using the
PmodBB breadboard module.

Author:        Dogan Ibrahim
Date:          May, 2014
File:          MELODY
Board:         chipKIT MX3
================================================================================*/
#define Max_Notes 25
//
// Tone variables
//
int buzzer = 0;                                            // buzzer port

//
// Melody frequencies
//
unsigned int Notes[Max_Notes] =
 {
    262, 262, 294, 262, 349, 330, 262, 262, 294, 262, 392,
    349, 262, 262, 524, 440, 349, 330, 294, 466, 466, 440,
    349, 392, 349
};

//
// Melody note durations
//
unsigned char Durations[Max_Notes] =
{
    1, 1, 2, 2, 2, 3, 1, 1, 2, 2, 2, 3, 1, 1, 2, 2, 2, 2, 2,
    1, 1, 2, 2, 2, 3
};
unsigned char i;


void setup()
{
        pinMode(buzzer, OUTPUT);                           // set buzzer as output
        digitalWrite(buzzer, LOW);                         // turn OFF buzzer to start with
}


void loop()
```

Figure 7.126: Program Listing *(Continued)*

```
{
    for(i = 0; i<Max_Notes; i++)                    // Do for all notes
    {
        tone(buzzer, Notes[i]);                     // Play the notes
        delay(400*Durations[i]);                    // durations
        noTone(buzzer);
        delay(100);                                 // Note gap
    }
    delay(3000);                                    // Repeat after 3 seconds
}
```

Figure 7.126: *(cont.)*

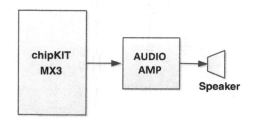

Figure 7.127: Block Diagram of the Project

Figure 7.128: The mikroElektronika Audio Amplifier Board

pair of screw-type connectors. The output of the amplifier board is connected to a speaker for high-volume and high-quality sound output.

The audio amplifier board uses an LM386-type audio amplifier chip, and its circuit diagram is shown in Figure 7.129.

Pmod connector power jumper is set to +5 V, and power is applied from pins JA-05 and JA-06 to the audio amplifier board. Pin JA-01 of the Pmod connector is connected to the audio input of the amplifier board. Figure 7.130 shows the hardware setup of the project.

The basic features of the audio amplifier board used in this project are as follows:

- LM386 audio power amplifier
- +5 V operation
- Easy connectivity using provided input and output screw terminals

The project PDL and project program are same as in Figures 7.125 and 7.126, respectively.

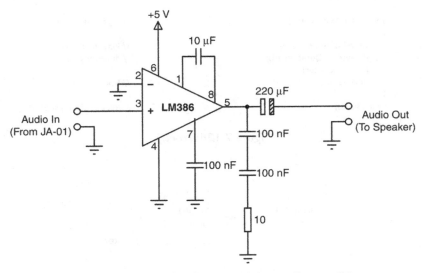

Figure 7.129: Circuit Diagram of the Audio Amplifier

Figure 7.130: Hardware Setup of the Project

7.27 Project 7.27 – Waveform Generation: Using DAC
7.27.1 Project Description

Waveform generation is very important in digital signal processing. In this project, we will be seeing how to generate simple waveforms using a digital-to-analogue converter (DAC) module.

Figure 7.131 shows the block diagram of a typical microcontroller-based waveform generation system. Here, the microcontroller generates the required waveform as a digital signal and then the DAC converts this signal into analogue. In practical applications, a low-pass filter is used after the DAC to clean the signal and remove any high-frequency components.

Basically two methods are used for waveform generation:

* The microcontroller calculates the waveform points in real time and sends them to the DAC.
* The waveform points are stored in a look-up table. The microcontroller reads these points from the table and sends them to the DAC (this method is used to generate any arbitrary waveform, or to generate higher-frequency waveforms).

As we shall see later, the rate at which the waveform points are sent to the DAC determines the frequency of the waveform.

Before going into the details of actually generating waveforms, it is worthwhile to look at the operation of a typical DAC. There are several forms of DACs available in the market. In this project, we will be using the R2R-type DAC that simply consists of resistors. The DAC we will be using in this project is the Pmod peripheral module PmodR2R, and it provides 8-bit conversion with up to 25 MHz operation. R2R-type DAC converters are cheap as they consist of only resistors and they are also fast.

Figure 7.132 shows a picture of the PmodR2R module. The connection diagram of the module is shown in Figure 7.133.

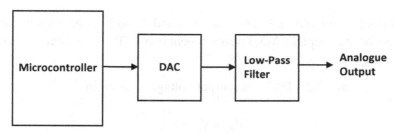

Figure 7.131: Block Diagram of Microcontroller-Based Waveform Generation

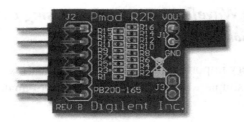

Figure 7.132: The PmodR2R DAC Module

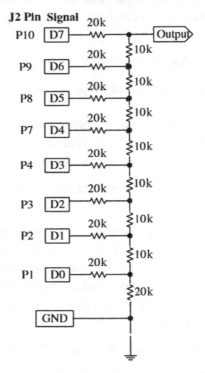

Figure 7.133: Connection Diagram of the PmodR2R Module

The PmodR2R module contains a 2 × 6 connector, and it can be connected to any of the Pmod connectors on the chipKIT MX3 development board. The converted analogue signal is available on header J1.

For a digital value D, of a R2R DAC, the output voltage is given by:

$$V_{\text{out}} = V_{\text{ref}} \times \frac{D}{2^N}$$

The PmodR2R is 8-bit wide and, therefore, $N = 8$, $2^N = 256$. With $V_{ref} = 3.3$ V (typical CMOS logic HIGH voltage), V_{out} will be:

$$V_{out} = 3.3 \times \frac{D}{2^8}$$

For $D = 1$, we have

$$V_{out} = 3.3 \times \frac{1}{256} = 12.89 \, \text{mV}$$

For $D = 255$, we have

$$V_{out} = 3.3 \times \frac{255}{256} = 3.28 \, \text{V}$$

Thus, the output voltage of the DAC changes from 0 to +3.28 V and the step size is 12.89 mV.

7.27.2 Project Hardware

The PmodR2R DAC module is connected to Pmod connector JA. Thus, the connection between the DAC module and the logical I/O ports is as follows:

PmodR2R Pin	Logical I/O Port	JA Connector Pin
D0	0	JA-01
D1	1	JA-02
D2	2	JA-03
D3	3	JA-04
D4	4	JA-07
D5	5	JA-08
D6	6	JA-09
D7	7	JA-10

Figure 7.134 shows the hardware setup. The circuit diagram of the project is shown in Figure 7.135.

Generating square waveform

In general, a square waveform with an amplitude equal to logic 1, that is, about +3.3 V (or +5 V) at audio frequencies can easily be generated by toggling an output pin of the microcontroller. In this section, we shall be using the DAC to generate a square waveform with an amplitude of +1.0 V and period of 10 ms, with a duty cycle of 50%, that is, 5 ms ON time and 5 ms OFF time.

Figure 7.134: Hardware Setup

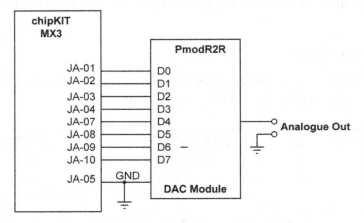

Figure 7.135: Circuit Diagram of the Project

The required program listing is shown in Figure 7.136. Logical port numbers 0–7 are configured as outputs. Then, for 1 V output, the data to be sent to the port is:

$$V_{out} = 3.3 \times \frac{D}{256}$$

or

$$D = 256 \times \frac{V_{out}}{3.3} = 256 \times \frac{1}{3.3}$$

or

$$D = 77$$

```
/*=============================================================================
                    GENERATING SQUARE WAVEFORM
                    ==============================

In this project the PmodR2R DAC module is used to generate a square waveform. The generated
waveform has the period of 10ms, duty cycle of 50% (5ms ON time, 5ms OFF time), and an
amplitude of 1.0V.

The output is sent to a digital oscilloscope so that it can be verified.


Author:         Dogan Ibrahim
Date:           May, 2014
File:           SQUARE
Board:          chipKIT MX3
=============================================================================*/
unsigned char PmodDAC[] ={0, 1, 2, 3, 4, 5, 6, 7};          // DAC logical port numbers
unsigned char Bits[8];
unsigned char i;


//
// This function separates a byte into 8 bits so that the byte can be sent to an 8-bit port easily
//
void port(unsigned char c)
{
        unsigned char d;

        for(i = 0; i< 8; i++)
        {
                d = 1 << i;
                if((c & d) != 0)
                        Bits[i] = 1;
                else
                        Bits[i] = 0;
        }
}
```

Figure 7.136: Program Listing *(Continued)*

Inside the main program loop, decimal number 77 is sent to the port and the port state is toggled every 5 ms so that the period of the square waveform is 10 ms. Array **PmodDAC** stores the logical I/O port numbers for the port at connector JA. These port pins are configured as output inside the *setup()* routine. In addition, the bit pattern of number 77 is found and stored in an array called **Bits**. Inside the main program loop, the bits are sent to the port pins for 5 ms. After this, the port pins are returned to 0 for 5 ms and this process is repeated, thus generating the required square wave. Function port converts a given byte into its bits and stores

```
void setup()
{
        unsigned char j;
        for(j = 0; j <= 7; j++)                          // do for all port pins
        {
                pinMode(PmodDAC[j], OUTPUT);              // set as outputs
        }
        port(77);                                        // convert to bits
}


void loop()
{
        for(i = 0; i< 8; i++)                            // ON time

        {
                digitalWrite(PmodDAC[i], Bits[i]);                 // send out
        }
        delay(5);                                        // wait 5ms

        for(i = 0; i< 8; i++)                            // OFF time
        {
                digitalWrite(PmodDAC[i], LOW);
        }
        delay(5);                                        // wait 5ms
}
```

Figure 7.136: *(cont.)*

these bits in array **Bits**, where **Bits[0]** corresponds to the LSB bit and **Bits[7]** corresponds to the MSB bit.

Figure 7.137 shows the generated waveform on a digital oscilloscope. In this book, the Velleman PCSGU250 PC-based digital oscilloscope is used to capture the images.

The program given in Figure 7.136 can be simplified by noticing that the pins JA-01 to JA-10 actually constitute PORTE of the microcontroller. Thus, we can send a byte directly to PORTE. The modified and simpler program is shown in Figure 7.138.

Generating sawtooth waveform

In this part of the project, we will be generating sawtooth waveform with the following speci-fications:

Output voltage: 0 to +3.3 V
Frequency: 100 Hz (period: 10 ms)
Step size: 0.1 ms

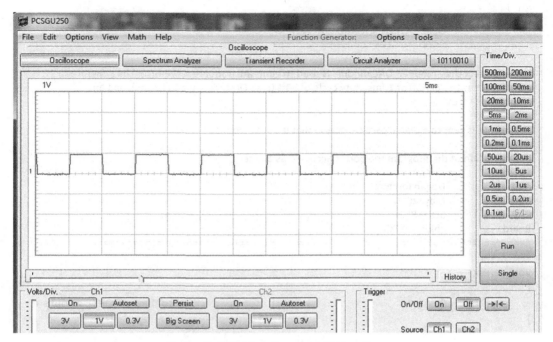

Figure 7.137: The Generated Waveform

The block diagram and the circuit diagram of the project are as in the previous project. The PmodR2R module is connected to Pmod connector JA as before.

The program is called SAWTOOTH, and the program listing is given in Figure 7.139. At the beginning of the program, PORTE is configured as output. Inside the main program loop, the steps of the sawtooth waveform are sent out. Since there are 10 steps in the waveform and the required frequency is 100 Hz, that is, period 10 ms, the duration of each step should be 10,000/10 = 1000 μs. The **delayMicroseconds**() function is used to generate the required delay in microseconds.

Figure 7.140 shows the output waveform obtained using the PSCGU250 digital oscilloscope. Here the vertical axis is 1 V/division and the horizontal axis is 5 ms/division. The graph is moved down the 0 V point for clarity.

Generating sine waveform

In this part of the project, we will see how to generate a low-frequency sine wave using the built-in trigonometric **sin** function, and then send the output to the DAC. The generated sine wave has amplitude of ±1 V, frequency of 50 Hz, and D.C. offset of 2.0 V.

The frequency of the sine wave to be generated is 50 Hz. This wave has a period of 20 ms, or 20,000 μs. If we assume that the sine wave will consist of 100 samples, then each sample

```
/*==================================================================================
                        GENERATING SQUARE WAVEFORM
                        ==============================
```

In this project the PmodR2R DAC module is used to generate a square waveform. The generated waveform has the period of 10ms, duty cycle of 50% (5ms ON time, 5ms OFF time), and an amplitude of 1.0V.

The output is sent to a digital oscilloscope so that it can be verified.

In this version of the program the port is accessed as a byte.

```
Author:          Dogan Ibrahim
Date:            May, 2014
File:            SQUARE2
Board:           chipKIT MX3
==================================================================================*/
```

```
void setup()
{
        TRISE = 0;                                      // configure PORTE as output
}



void loop()
{
        PORTE = 77;                                     // send 77 to PORTE
        delay(5);                                       // wait 5ms
        PORTE = 0;                                       // Send 0 to PORTE
        delay(5);                                       // wait 5ms
}
```

Figure 7.138: Modified Simpler Program

should be output at $20{,}000/100 = 200$ μs intervals. The sample values will be calculated using the trigonometric **sin** function of the compiler.

The sin function will have the following format:

$$\sin\left(\frac{2\pi \times \text{Count}}{T}\right)$$

where T is the period of the waveform and is equal to 100 samples. Count is a variable that ranges from 0 to 100 and is incremented by 1 at each iteration. Thus, the sine wave is

```
/*=====================================================================
                    GENERATING SAWTOOTH WAVEFORM
                    ===============================
```

In this project the PmodR2R DAC module is used to generate a sawtooth waveform. The generated waveform has the following specifications:

Output voltage: 0 to +3.3V
Frequency: 100Hz (period: 10ms)
Step size: 0.1ms

The output is sent to a digital oscilloscope so that it can be verified.

In this version of the program the port is accessed as a byte.

Author: Dogan Ibrahim
Date: May, 2014
File: SAWTOOTH
Board: chipKIT MX3
```
=====================================================================*/

void setup()
{
        TRISE = 0;                                          // configure PORTE as output
}



void loop()
{
        for(floati = 0; i<= 1; i = i + 0.1)                 // send out sawtooth steps
        {
                PORTE = i*255;
                delayMicroseconds(1000);                    // 1000us delay
        }
}
```

Figure 7.139: Program Listing

divided into 100 samples and each sample is output at 200 μs. The above formula can be rewritten as:

$$\sin(0.0628 \times \text{Count})$$

It is required that the amplitude of the waveform should be ±1 V. With a reference voltage of +3.3 V and an 8-bit R2R DAC converter, 1 V is equal to decimal number 77

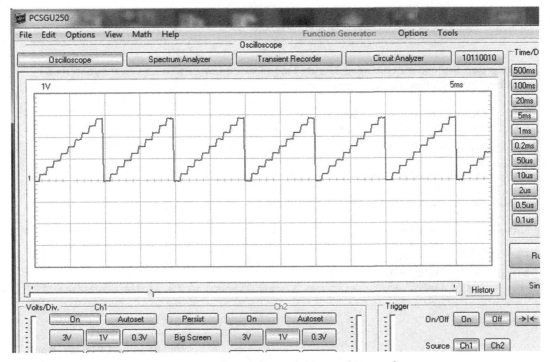

Figure 7.140: The Generated Sawtooth Waveform

(see previous project). Thus, we will multiply our sine function with the amplitude at each sample to give:

$$77 \times \sin(0.0628 \times \text{Count})$$

The DAC converter used in this project is unipolar and cannot output negative values. Therefore, a D.C. offset is added to the sine wave to shift it so that it is always positive. The offset should be larger than the absolute value of the maximum negative value of the sine wave, which is -77 when the **sin** function above is equal to -1. In this project, we wish to add 2.0 V D.C. offset, which corresponds to a decimal value of 155 at the DAC output. Thus, at each sample, we will calculate and output the following value to the DAC:

$$155 + 77 \times \sin(0.0628 \times \text{Count})$$

The program is called SINE, and its listing is shown in Figure 7.141. The sine wave amplitude is set to 77, offset is set to 155, and variable R is defined as $2\pi/100$. The sine waveform values for a period are obtained in the *setup()* routine outside the main program loop. The reason for calculating these values outside the main program loop is to minimise the loop time so that higher-frequency sine waves can be generated. The sine values are calculated as follows:

```
for(i = 0; i < 100; i++)sins[i] = offset + Amplitude*sin(R*i);
```

```
/*=================================================================================
                            GENERATING SINE WAVEFORM
                            ===========================

In this project the PmodR2R DAC module is used to generate a sine waveform. The generated
waveform has the following specifications:

        Output voltage:   ±1.0V
        Output offset:    2.0V
        Frequency:        50Hz (period: 20ms)

The output is sent to a digital oscilloscope so that it can be verified.

In this version of the program the port is accessed as a byte.

Author:        Dogan Ibrahim
Date:          May, 2014
File:          SINE
Board:         chipKIT MX3
=================================================================================*/
#define T 100                                      // 100 samples
#define R 0.0628                                    // 2*PI/T
#define Amplitude 77                                // amplitude = 1V
#define offset 155                                  //offset = 2.0V
float sins[100];                                    // sine values
unsigned char Count = 0;
unsignedint Value;


void setup()
{
        TRISE = 0;                                  // configure PORTE as output
//
// Generate the sine wave samples offline and load into an array called sins
//
        for(inti = 0; i< 100; i++)sins[i] = offset+Amplitude*sin(R*i);
//

}




void loop()
{
//
// Get sine wave samples and send to DAC
//
        Value = sins[Count];                        // get a sine sample
         PORTE = Value;                             // Send to DAC converter
        Count++;                                    // increment Count
        if(Count == 100)Count = 0;
        delayMicroseconds(200);                     // wait 200us
}
```

Figure 7.141: Program Listing

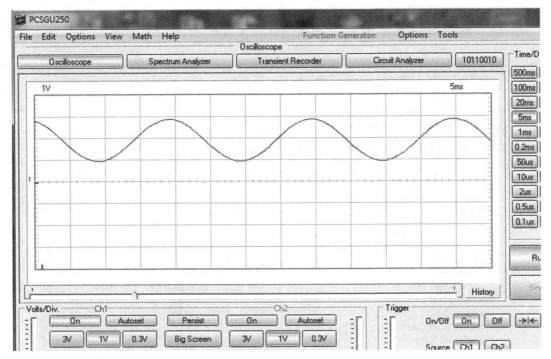

Figure 7.142: Generated Waveform

Inside the main program, the sine samples are output with 200 μs delays so that the waveform has the required frequency.

Figure 7.142 shows the waveform generated by the program. It is clear from this figure that the generated sine waveform has period 20 ms as designed. Here, the vertical axis is 1 V/division and the horizontal axis is 5 ms/division.

7.28 Project 7.28 – Using a Keypad

7.28.1 Project Description

Keypads are small keyboards that are used to enter numeric or alphanumeric data to microcontroller systems. Keypads are available in a variety of sizes and styles from 2 × 2 to 4 × 4 or even bigger.

In this project, the PmodKYPD module is used. This is a 4 × 4 keypad with 2 × 6 pins and keys labelled 0–9 and A–F. Figure 7.143 shows a picture of the PmodKYPD module.

This is a simple project where the Serial Monitor displays the key pressed on the keypad.

Figure 7.143: The PmodKYPD Module

7.28.2 Keypad Structure

Figure 7.144 shows the structure of a typical 4 × 4 keypad, consisting of 16 switches, formed in a 4 × 4 array. Assuming the keypad is connected to PORT C, the steps to detect which key is pressed are as follows:

- A logic 1 is applied to first column via RC0.
- Port pins RC4–RC7 are read. If the data is nonzero, then a switch is pressed. If RC4 is 1, key 1 is pressed; if RC5 is 1, key 4 is pressed; if RC6 is 1, key 9 is pressed; and so on.
- A logic 1 is applied to second column via RC1.

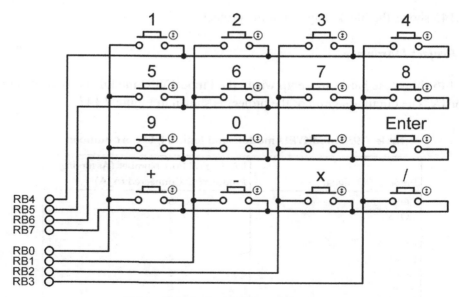

Figure 7.144: 4 × 4 Keypad Structure

Table 7.8: PmodKYPD pin layout.

Connector J1 – Column/Row Indicators		
Pin	Signal	Description
1	COL4	Column 4
2	COL3	Column 3
3	COL2	Column 2
4	COL1	Column 1
5	GND	Power supply ground
6	VCC	Power supply (3.3 V)
7	ROW4	ROW4
8	ROW3	ROW3
9	ROW2	ROW2
10	ROW1	ROW1
11	GND	Power supply ground
12	VCC	Power supply (3.3 V)

- Again port pins RC4–RC7 are read. If the data is nonzero, then a switch is pressed. If RC4 is 1, key 2 is pressed; if RC5 is 1, key 6 is pressed; if RC6 is 1, key 0 is pressed; and so on.
- The above process is repeated for all four columns continuously.

Table 7.8 shows the pin layout of the PmodKYPD module. Assuming that the keypad is connected to Pmod connector JA, the column and row numbers and corresponding I/O port logical names are shown in Table 7.9.

Figure 7.145 shows the block diagram of the project.

7.28.3 Project Hardware

Figure 7.146 shows the hardware setup where the PmodKYPD module is connected to Pmod connector JA. The circuit diagram of the project is shown in Figure 7.147.

Table 7.9: PmodKYPD module and logical I/O port numbers.

PmodKYPD Signal	I/O Logical Port Number (Assuming PmodKYPD Connected to JA)
COL4	0
COL3	1
COL2	2
COL1	3
ROW4	4
ROW3	5
ROW2	6
ROW1	7

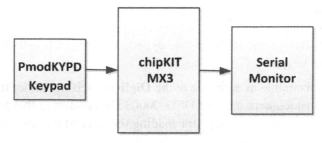

Figure 7.145: Block Diagram of the Project

Figure 7.146: Project Hardware Setup

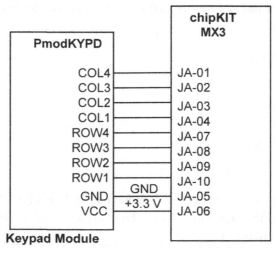

Figure 7.147: Circuit Diagram of the Project

7.28.4 Project PDL

Before going into details of programming the keypad, it is worthwhile to look at the functions available for using the keypad.

The keypad library is available as a zip file at the **Digilent** web site under the PmodKYPD module specifications (document number DSD-0000337). Assuming that you have installed MPIDE in the default directory, the steps for loading the keypad library are given as follows:

- Copy the keypad library to a folder and unzip it. You should have the following files:
    KYPD: C++ source file
    KYPD: header file
    keywords: text file
    documents: folder (keypad documentation)
    examples: folder (keypad example)
- Create a new folder with the name KYPD under the following subdirectory (assuming you have the same version of MPIDE) and copy all the unzipped files into this directory:
    C:\MPIDE\mpide-0023-windows-20130715\hardware\pic32\libraries
- Start the MPIDE. Click *Sketch → Import Library → KYPD*.

You should see the header file **KYPD.h** copied to the beginning of your file.

After including the header file, we have to define the number of rows and columns on our keypad. For a 4 × 4 keypad, this can be done as follows:

```
const byte ROWS = 4;
const byte COLS = 4;
```

Then, we have to specify the structure of our keypad. For the 4 × 4 keypad that we will be using, this can be done as follows (notice that the characters on the keyboard matrix can be changed if desired):

```
int keys[COLS][ROWS] =
{
        {1, 4, 7, 0},
        {2, 5, 8, 'F'},
        {3, 6, 9, 'E'},
        {'A', 'B', 'C', 'D'}
};
```

We should now define the interface between the keypad and the chipKIT MX3 development board. In reference to Table 7.9, we can define the interface rows and columns as follows, starting from row 1 and column 1:

```
unsigned int RowPins[ROWS] = {7, 6, 5, 4};
unsigned int ColPins[COLS] = {3, 2, 1, 0};
```

Then, a keypad object should be created from the Keypad class as follows:

KYPD mykpd;

We can now use object **mykpd** with the various functions of the keypad library. A brief summary of the commonly used functions is given as follows:

mykpd.begin(): This function sets column pins for outputs and row pins for inputs.
mykpd.setPins(rows, cols): This function sets the pins for rows and columns.
mykpd.setKeyMap(keys): This function maps the key table.
mykpd.end(): This function releases the pins.
mykpd.getKey(): This function returns the key that is pressed (if any). The function is nonblocking. Returns −1 if no keys were pressed.
mykpd.getColRow(): This function returns 32-bit result. The column is returned in the most significant 16-bit, and the row in the least significant 16-bit. Returns −1 if no keys were pressed.
mykpd.getKey(colrow): In this function, **colrow** is the column–row indicator. The function returns the key pressed.

We can find the key pressed by either calling the **getKey**() function directly or calling the **getColRow**() function to find the column and row indicator and then use this value in the call **getKey(colrow)** to find the actual key pressed.

The project PDL is shown in Figure 7.148.

7.28.5 Project Program

The program is called KEYPAD1, and the program listing is given in Figure 7.149. At the beginning of the program, the keypad library **KYPD.h** is declared. Then, the structure and the interface of the keypad are defined in the *setup()* routine and also the Serial Monitor is

```
BEGIN
        Define rows, columns, and keypad map
        Define keypad I/O ports
        Configure Serial Monitor
        Configure keypad
        DO FOREVER
                Get a key
                IF a key is pressed THEN
                        Display key on Serial Monitor
                ENDIF
                Wait 200ms
        ENDDO
END
```

Figure 7.148: Project PDL

```
/*================================================================================
                        KEYPAD EXAMPLE
                        ===============

In this project the PmodKYPD 4x4 keypad module is connected to Pmod connector JA.

The program checks if a key is pressed on the keyboard and then displays this key on the
Serial Monitor.


Author:         Dogan Ibrahim
Date:           May, 2014
File:           KEYPAD1
Board:          chipKIT  MX3
================================================================================*/
#include <KYPD.h>                                    // keypad library

const byte ROWS = 4;                                 // no of rows
const byte COLS = 4;                                 // no of columns

int keys[COLS][ROWS] =                               // keypad map
{
        {1, 4, 7, 0},
        {2, 5, 8, 'F'},
        {3, 6, 9, 'E'},
        {'A', 'B', 'C', 'D'}
};

unsignedintRowPins[ROWS] = {7, 6, 5, 4};             // row I/O ports
unsignedintColPins[COLS] = {3, 2, 1, 0};             // column I/O ports
int c;
KYPD mykpd;                                          // object mykpd

void setup()
{
        Serial.begin(9600);                          // initialize Serial Monitor
        mykpd.setPins(RowPins, ColPins);             // setPins
        mykpd.setKeyMap(keys);                       // map keypad
        mykpd.begin();                               // begin
}



void loop()
{
        c = mykpd.getKey();                          // get a key
        if(c != -1)                                  // if a key is pressed
        {
                Serial.println(c);                   // display the key
        }
        delay(200);                                  // wait a bit
}
```

Figure 7.149: Program Listing

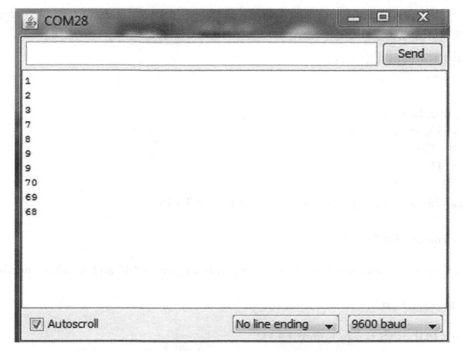

Figure 7.150: Example Output on the Serial Monitor

initialised at 9600 Baud. Inside the main program loop, the program checks if a key is pressed and if so displays the pressed key on the Serial Monitor.

Figure 7.150 shows an example output from the program where the following keys were pressed on the keypad: **1237899FED**.

7.29 Project 7.29 – Keypad Calculator
7.29.1 Project Description

In this project, we design a simple four-function (+, −, ×, /) calculator using the PmodKYPD keypad and the Serial Monitor.

The user first enters two numbers and then the operation to be performed. The result is calculated and displayed on the Serial Monitor.

In this project, the keypad is organised as follows:

```
1  2  3  +
4  5  6  -
7  8  9  *
0  F  E  /
```

where **E** is the entry (return) key. The **F** key is not used. The operation of the calculator is shown as follows for multiplying two numbers as an example:

SIMPLE CALCULATOR
====================

First number   : 3 <E>
Second number: 5 <E>
Op: * <E>
Res = 15

The block diagram of the project is same as in Figure 7.145.

7.29.2 Project Hardware

The hardware setup and the circuit diagram are as in Figures 7.146 and 7.147, respectively.

7.29.3 Project PDL

The project PDL is shown in Figure 7.151.

7.29.4 Project Program

The program is called CALC, and its listing is shown in Figure 7.152. At the beginning of the program, the keypad library **KYPD.h** is included and various keypad keys are defined. The keypad map is stored in an array called **keys**. Inside the *setup()* routine, the Serial Monitor is initialised and the keypad library is mapped to the I/O ports.

```
BEGIN
        Include keypad library
        Define various keypad constants and variables
        Configure Serial Monitor
        DO FOREVER
                Display heading
                Display First number  :
                Read first number
                Display Second number :
                Read second number
                Display Op:
                Read operation
                Perform operation
                Display result (Res = ) on Serial Monitor
        ENDDO
END
```

Figure 7.151: Project PDL

```
/*============================================================================
                        KEYPAD BASED CALCULATOR
                        ========================

In this project the PmodKYPD 4x4 keypad module is connected to Pmod connector JA.

The program is a simple 4-function calculator (+ - * /). The data entry and results are displayed on
the Serial Monitor. An example output is shown below:

        SIMPLE CALCULATOR
        =================

        First number : 5
        Second number: 3
        Op: +
        Res = 8

Author:         Dogan Ibrahim
Date:           May, 2014
File:           CALC
Board:          chipKIT MX3
============================================================================*/
#include <KYPD.h>                                      // keypad library

#define Enter 'E'                                       // ENTER key
#define Plus '+'                                        // + key
#define Minus '-'                                       // - key
#define Multiply '*'                                    // * key
#define Divide '/'                                      // / key
unsigned char i,j,op[12];
unsigned long Calc, Op1, Op2;
//
// Keypad constants and variables
//
const byte ROWS = 4;                                    // no of rows
const byte COLS = 4;                                    // no of columns

int keys[COLS][ROWS] =                                  // keypad map
{
        {1,   4,   7,   0},
        {2,   5,   8,  'F'},
        {3,   6,   9,  'E'},
        {'+', '-', '*', '/'}
};
unsignedintRowPins[ROWS] = {7, 6, 5, 4};                // row I/O ports
unsignedintColPins[COLS] = {3, 2, 1, 0};                // column I/O ports
int c, Mykey;
KYPD mykpd;                                             // object mykpd

void setup()
{
        Serial.begin(9600);                             // initialize Serial Monitor
        mykpd.setPins(RowPins, ColPins);                // setPins
        mykpd.setKeyMap(keys);                          // map keypad
        mykpd.begin();                                  // begin
```

Figure 7.152: Program Listing *(Continued)*

```
        }


void loop()
{
//
// Display heading
//
        Op1 = 0;
        Op2 = 0;
        Serial.println("");
        Serial.println("SIMPLE CALCULATOR");
        Serial.println("=================");
        Serial.println("");
//
// Get first number
//
        Serial.print("First number : ");
        while(1)
        {
                do
        {
                delay(100);
                        Mykey = mykpd.getKey();         // get a key
                }while(Mykey == -1);                    // if no key pressed
                if(Mykey == Enter)break;                // if ENTER pressed
                if(Mykey == '0')Mykey = 0;              // if 0 is pressed
                Serial.print(Mykey);                    // display
                Op1 = 10*Op1 + Mykey;                   // entered number
        delay(200);
        }
        Serial.println("");

//
// Get second number
//
        Serial.print("Second number: ");
        while(1)
        {
                do
        {
                delay(100);
                        Mykey = mykpd.getKey();         // get a key
                }while(Mykey == -1);                    // if no key pressed
                if(Mykey == Enter)break;                // if ENTER pressed
                if(Mykey == '0')Mykey = 0;              // if 0 is pressed
                Serial.print(Mykey);                    // display
                Op2 = 10*Op2 + Mykey;                   // entered number
        delay(200);
        }
        Serial.println("");

//
// Get operation
//
        Serial.print("Op: ");
```

Figure 7.152: *(cont.)*

```
        do
        {
                delay(100);
                Mykey = mykpd.getKey();          // get  a key
        }while(Mykey == -1);                     // if no key pressed
        Serial.write(Mykey);                     // display key
        Serial.println("");
        Serial.print("Res = ");
//
// Do the required operation
//
    switch(Mykey)                                // Perform the operation
    {
    case Plus:
        Calc = Op1 + Op2;                        // If ADD
         break;
    case Minus:
        Calc = Op1 - Op2;                        // If Subtract
        break;
    case Multiply:
        Calc = Op1 * Op2;                        // If Multiply
        break;
    case Divide:
        Calc = Op1 / Op2;                        // If Divide
        break;
    }
//
// Display the result
//
    Serial.println(Calc);
    Serial.println("");
}
```

Figure 7.152: *(cont.)*

Inside the main program loop, the first number, the second number, and the operation to be performed are read. The numbers are read using the following code:

```
Serial.print("First number : ");
while(1)
{
        do
        {
                delay(100);
                Mykey = mykpd.getKey();          // get a key
        }while(Mykey == -1);                     // if no key pressed
        if(Mykey == Enter)break;                 // if ENTER pressed
        if(Mykey == '0')Mykey = 0;               // if 0 is pressed
        Serial.print(Mykey);                     // display
        Op1 = 10*Op1 + Mykey;                    // entered number
        delay(200);
}
Serial.println("");
```

Keypad library routine **mykpd.getKey**() is used in a **while** loop until a key is pressed by the user. Notice that a small delay is used inside this loop to avoid multiple key presses. The program exits the while loop when the **Enter** key is pressed. Otherwise, the entered numbers are collected and the total number is stored in a variable called **Op1** (or **Op2** for the second number).

A **switch** loop is used to perform the required operation:

```
switch(Mykey)                                    // Perform the operation
{
  case Plus:
      Calc = Op1 + Op2;                          // If ADD
      break;
  case Minus:
      Calc = Op1 - Op2;                          // If Subtract
      break;
  case Multiply:
      Calc = Op1 * Op2;                          // If Multiply
      break;
  case Divide:
      Calc = Op1 / Op2;                          // If Divide
      break;
}
```

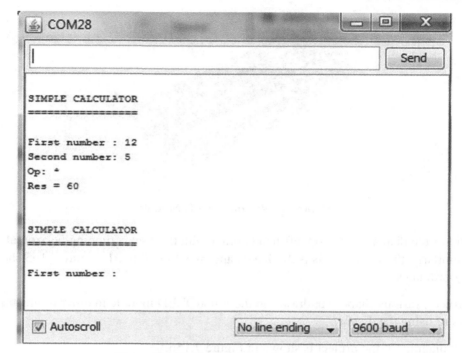

Figure 7.153: Typical Run of the Program

The result is in variable **Calc** and is displayed on the Serial Monitor.

A typical run of the program is shown in Figure 7.153.

7.30 Project 7.30 – Using Graphics LCD

7.30.1 Project Description

Graphics LCDs (GLCDs) are very useful in many embedded applications, requiring visual graphical outputs. There are many types of GLCDs such as monochrome, colour, TFT, organic (OLED), etc.

Two sizes of GLCDs are available as Pmod modules: the small-size PmodOLED and the bigger-size PmodOLED2. In this project, the smaller PmodOLED is used.

The PmodOLED is an organic 128×32 pixel GLCD having size 0.9 inches. The display uses the Solomon Systech SSD1306 display controller and is controlled from the standard SPI bus interface. The display module is model UG2832, and it operates at clock rates up to 10 MHz and is connected to the chipKIT MX3 development board through a 12-pin connector.

Figure 7.154 shows a picture of the PmodOLED module.

Figure 7.154: PmodOLED Module

The origin of the display is the top left-hand corner with the *X*-axis running horizontal from 0 to 127. Similarly, the vertical axis is the *Y*-axis and runs from 0 to 31. Figure 7.155 shows the display coordinates.

In this project, various shapes are drawn on the PmodOLED module in order to illustrate the principles of using the display.

The block diagram of the project is shown in Figure 7.156.

7.30.2 Project Hardware

The PmodOLED display is connected to Pmod connector JE where the SPI bus is located. The pin configuration of the display is shown in Table 7.10. Figure 7.157 shows the hardware setup. The circuit diagram of the project is shown in Figure 7.158.

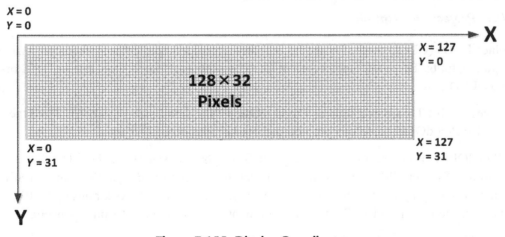

Figure 7.155: Display Coordinates

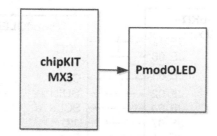

Figure 7.156: Block Diagram of the Project

Table 7.10: PmodOLED pin configuration

Connector J1		
Pin	Signal	Description
1	CS	SPI chip select (slave select)
2	SDIN	SPI data in (MOSI)
3	None	Unused pin
4	SCLK	SPI clock
7	D/C	Data/command control
8	RES	Power reset
9	VBATC	V_{BAT} battery voltage control
10	VDDC	V_{DD} logic voltage control
5, 11	GND	Power supply ground
6, 12	VCC	Power supply

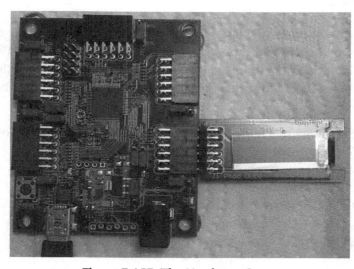

Figure 7.157: The Hardware Setup

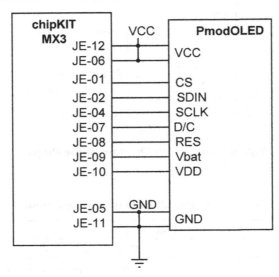

Figure 7.158: Circuit Diagram of the Project

The PmodOLED module has two power supplies: VDD is the power to the display logic, and Vbat is the power to the actual display. The display/control (D/C) pin determines whether the bytes sent to the display should be interpreted as data or commands. The RES pin when driven LOW resets the display, and must be set HIGH for normal operation. The CS, SDIN, and SCLK are the standard SPI bus interface signals.

7.30.3 Project PDL

The PmodOLED library provides a large number of functions for using the display. This library is available in the **Digilent** PmodOLED product page (DSD-0000319) and should be installed into the MPIDE environment before it can be used. Assuming that you have installed MPIDE in the default directory, the steps for loading the OLED library are given as follows:

• Copy the OLED library to a folder and unzip it. You should have the following files:
  OLED: C++ source file
  OLED: header file
  keywords: text file
  documents: folder (keypad documentation)
  examples: folder (keypad example)
  Some other files

- Create a new folder with the name OLED under the following subdirectory (assuming you have the same version of MPIDE) and copy all the unzipped files into this directory:
  C:\MPIDE\mpide-0023-windows-20130715\hardware\pic32\libraries
- Start the MPIDE. Click *Sketch → Import Library → OLED*.

The following header files must be included at the beginning of your file:

```
#include <OLED.h>
#include <DSPI.h>
```

The display memory is organised such that the first byte (byte 0) corresponds to a vertical column of eight pixels at the left side of the display, with the least significant bit the uppermost pixel in the column. The next byte corresponds to the next column of eight pixels to the right of the first. This continues across the display to byte 127, which is the rightmost column of eight pixels. Byte 128 corresponds to the next column of eight pixels at the left of the display.

The display can operate in graphics or in character mode. The character can be selected. With the standard character size = 1, the character set is 5×7 dot matrix. With one pixel margin at the right and one pixel margin at the bottom, the character size is 6×8 pixels. The selected size value multiplies the height and width by the size. For example, with size = 2, the character will be $(2 \times 6) \times (2 \times 8) = 12 \times 16$ pixels. With 128×32 pixels and size = 1, there can be up to 4 lines of text and each line can have $128 / 6 = 21$ characters across, that is, the character display size is 4 rows $\times$ 21 columns (the character coordinates are 0 based, that is, 0–3 rows and 0–20 columns). With size = 2, there can be up to 2 lines of text and each line can have $128/12 = 10$ characters across, that is, the character display size is 2 rows $\times$ 10 columns (0–1 row and 0–9 columns).

A brief summary of the commonly used OLED functions is given below.

General functions

OLED.begin(): This function initialises the display and turns power to the display. It must be called before calling any other OLED function.

OLED.end(): This function turns power off to the display and frees the SPI pins used for the display.

OLED.displayON():This function activates the display.

OLED.displayOFF():This function blanks the display.

OLED.clear():This function clears the memory buffer and hence the display.

Character functions

OLED.setCursor(horz, vert): This function sets the horizontal (x) and vertical (y) character positions to the specified location.

OLED.getCursor(horz, vert): This function returns the horizontal (x) and vertical (y) character positions.

OLED.setCharUpdate(f): This function enables (nonzero) or disables (zero) automatic character update.

OLED.putchar(c): This function writes the specified character at the current cursor position and advances the cursor.

OLED.putString(str): This function writes the specified NULL terminated string to the display and advances the cursor.

Graphics functions

OLED.setDrawColor(clr): This function sets the foreground colour used for pixel draw.

OLED.getstdPattern(pat): This function returns a pointer for the specified pattern number.

OLED.setFillPattern(pat): This function sets a pointer to a fill pattern. There are eight fill patterns to choose from.

OLED.moveTo(x, y): This function sets the current graphics drawing coordinate to x, y.

OLED.getPos(x, y): This function returns the current graphics drawing coordinate.

OLED.drawPixel(): This function sets the pixel at the current drawing position.

OLED.drawLine(x, y): This function draws a line from the current pixel position to the specified position. Cursor is set at the new position.

OLED.drawRect(x, y): This function draws a rectangle bounded with the current location and the specified corners.

OLED.drawFillRect(x, y): This function fills a rectangle without drawing an outline around the rectangle.

OLED.putBmp(x, y, bmp): This function draws the specified bitmap.

OLED.drawChar(c): This function draws the specified character at the current location.

OLED.drawstring(str): This function writes the specified NULL terminated string to the display.

The PDL of the program is shown in Figure 7.159. The program displays the following strings and shapes:

"chipKIT"
"MX3"
"Development"
"Kit"
Rectangle at centre of display
Rectangle at centre of display with eight different fill patterns
X–Y-axis at the centre of display
X–Y-axis at bottom left of the display

```
BEGIN

        Include SPI and OLED libraries
        Power OLED module
        Display "chipKIT" on first row
        Display "MX3" on second row
        Display "Development" on third row
        Display "Kit" on fourth row
        Wait 5 seconds
        Blink display 3 times with 1 second intervals
        Draw a rectangle at center of the display
        Wait 5 seconds
        DO 8 TIMES
                Draw a rectangle at center of display
                Fill the rectangle with a pattern
                Wait 2 seconds
        ENDDO
        Wait 5 seconds
        Draw X-Y axis at the center of the display
        Wait 5 seconds
        Draw X-Y axis starting at the bottom left of the display
        Wait 5 seconds
END
```

Figure 7.159: Program PDL

7.30.4 Project Program

The program is named OLED1, and its listing is given in Figure 7.160. At the beginning of the program, the SPI library and the OLED library are included in the program. In the *setup( )* routine, the OLED is initialised and power applied to the module. Inside the main program loop, the character cursor is positioned to the left-hand side of the display and the following text is displayed:

Row 1: "chipKIT"
Row 2: "MX3"
Row 3: "Development"
Row 4: "Kit"

The program then waits for 5 s and blinks the display ON and OFF three times with 1-s delay between each blinking. After 5-s delay, the program displays a rectangle at the centre of the display. The graphics cursor is first set at coordinate (55, 1), which is the top left corner of the rectangle. The bottom right-hand corner of the rectangle is then set to (75, 27). The rectangle is displayed for 5 s. The program then establishes a *for* loop that executes eight times. Inside this loop, a rectangle is drawn and is filled with a different pattern at each iteration (pattern

```
/*==================================================================================
                        USING THE OLED DISPLAY
                        ======================
```

In this project the PmodOLEDgraphics LCD module is used. The module is connected to Pmod connector JPE where the SPI bus is.

PmodOLED is a 128 x 32 pixels, 0.9 inch small graphics display.

The program writes various characters on the display and also draws various shapes. Delays are inserted between outputs so that the user can make reference to the codes for displayed shapes.

```
Author:           Dogan Ibrahim
Date:             May, 2014
File:             OLED1
Board:            chipKIT MX3
=================================================================================*/

#include <DSPI.h>                                  // include SPI library
#include <OLED.h>                                  // include OLED library

OledClass OLED;                                     // define OLED object

void setup()
{
  OLED.begin();                                     // initialize & power ON OLED
}

void loop()
{
        intirow;
        int fill;
//
// Display text as follows:
//
// First row : chipKIT
// Second row: MX3
// Third row : Development
// Fourt row : Kit
//
        OLED.clearBuffer();                         // clear memory
        OLED.setCursor(0, 0);                       // cursor at character pos 0,0
        OLED.putString("chipKIT");                  // write "chipKIT"
        OLED.setCursor(0, 1);                       // cursor at character pos 0,1
        OLED.putString("MX3");                      // write "MX3"
        OLED.setCursor(0, 2);                       // cursor at character pos 0,2
        OLED.putString("Development");              // write "Development"
        OLED.setCursor(0,3);                        // cursor at character pos 0,3
        OLED.putString("Kit");                      // write "Kit"
        delay(5000);                                // wait 5 seconds
//
// Blink the display several times
//
        OLED.displayOff();                          // display OFF
```

Figure 7.160: Program Listing *(Continued)*

```
            delay(1000);                                  // wait 1 second
            OLED.displayOn();                             // display ON
            delay(1000);                                  // wait 1 second
            OLED.displayOff();                            // display OFF
            delay(1000);                                  // wait 1 second
            OLED.displayOn();                             // display ON
            delay(1000);                                  // wait 1 second
            OLED.displayOff();                            // display OFF
            delay(1000);                                  // wait 1 second
            OLED.displayOn();                             // display ON
            delay(5000);                                  // wait 5 seconds
//
// Clear the display
//
            OLED.clear();                                 // clear display
//
// Draw a rectangle at center of the screen
//
            OLED.moveTo(55, 1);                           // go to 55,1
            OLED.drawRect(75, 27);                        // draw rectangle at 55,1 to 75,27
            OLED.updateDisplay();                         // update display
            delay(5000);                                  // wait 5 seconds
//
// Draw a rectangle at centre again and fill with different patterns
//
            for(fill = 0; fill < 8; fill++)               // do for all 8 fill patterns
            {
                    OLED.clearBuffer();                   // clear buffer
                    OLED.setFillPattern(OLED.getStdPattern(fill));
                    OLED.moveTo(55, 1);                   // go to 55,1
                    OLED.drawFillRect(75, 27);            // fill rectangle
                    OLED.drawRect(75, 27);                // draw rectangle
                    OLED.updateDisplay();                 // update display
                    delay(2000);                          // wait 2 seconds
            }
            delay(5000);                                  // wait 5 seconds
//
// Clear display and draw an X-Y axis with origin in the middle of screen
//
            OLED.clear();                                 // clear display
            OLED.moveTo(64, 31);                          // go to 64,31
            OLED.drawLine(64, 0);                         // draw line to 64,0
            OLED.moveTo(0, 16);                           // go to 0,16
            OLED.drawLine(127, 16);                       // draw line to 127,16
            OLED.updateDisplay();                         // update display
            delay(5000);                                  // wait 5 seconds
//
// Clear display and draw an X-Y axis with center at bottom left
//
            OLED.clear();                                 // clear display
            OLED.moveTo(0, 31);                           // go to 0,31
            OLED.drawLine(127, 31);                       // draw line to 127,31
            OLED.moveTo(0, 31);                           // go to 0,31
            OLED.drawLine(0, 0);                          // draw line to 0,0
            OLED.updateDisplay();                         // update display
delay(5000);                                              // wait 5 seconds
}
```

Figure 7.160: *(cont.)*

0 is with no filling). Each pattern is displayed for 2 s. After 5-s delay, the program draws an X–Y-axis where the origin is at the centre of the display. Finally, another X–Y-axis is drawn with the centre at the bottom left corner of the display.

Figures 7.161–7.165 show the outputs on the display.

7.31 Project 7.31 – Creating an Image on the OLED

7.31.1 Project Description

In some applications, we may want to create and display our own images (e.g., a logo) on the OLED. In this section, the steps are given for creating and displaying a simple image.

The block diagram, hardware setup, and the circuit diagram of the project are as in Figures 7.156–7.158, respectively. PmodOLED is connected to Pmod connector JE as in the previous project.

Figure 7.161: Displaying the Text

Figure 7.162: Displaying a Rectangle

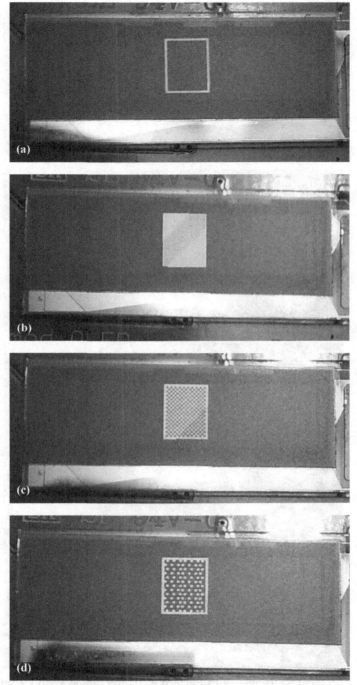

Figure 7.163: (a) Rectangle With Pattern 0. (b) Rectangle With Pattern 1. (c) Rectangle With Pattern 2. (d) Rectangle With Pattern 3. *(Continued)*

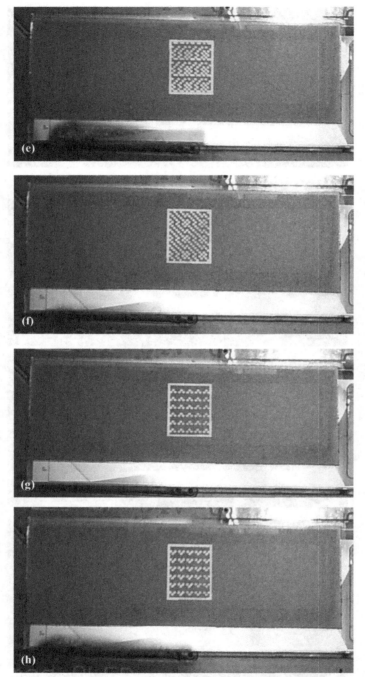

Figure 7.163: (e) Rectangle With Pattern 4. (f) Rectangle With Pattern 5.
(g) Rectangle With Pattern 6. (h) Rectangle With Pattern 7. *(cont.)*

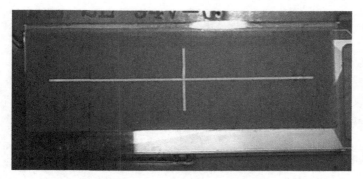

Figure 7.164: *X–Y*-Axis at the Centre of the Display

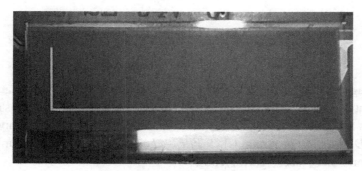

Figure 7.165: *X–Y*-Axis at the Bottom Left Corner of the Display

Creating the image

The image must be in bitmap (.bmp) format with the pixel size 128 × 32. There are many programs that can be helpful for creating bitmap images. Perhaps the easiest one is the **Paint** program, which is distributed free of charge with the Windows operating system. The steps for creating an image are given as follows:

- Start the Paint program (usually in folder *Accessories*).
- Click left-hand menu and select *Properties*.
- Set the width to 128 and height to 32 pixels (see Figure 7.166).
- Click *Zoom in* icon until the size of the bitmap is big enough.
- Draw your image using the cursor. In this example, the image consists of four rectangles as shown in Figure 7.167.
- Save the image as a BMP picture by clicking the *Save As* button and giving a name to the image.

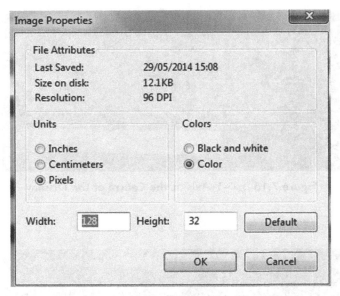

Figure 7.166: Select the Width and Height

- Now, the image should be converted to an array of bitmap values. There are several programs that can be used for this purpose. The one used in this project is called *LCD Assistant* and is available at the web site: http://en.radzio.dxp.pl/bitmap_converter.
- Start the *LCD Assistant* program. Click *File → Load image* and load your bitmap image created earlier. Select (see Figure 7.168):
    Byte orientation = vertical
    Width = 128, height = 32
    Size endianness = little
    Pixels/byte = 8
    and give a name to your output file (in this example, the name BOXES is given).

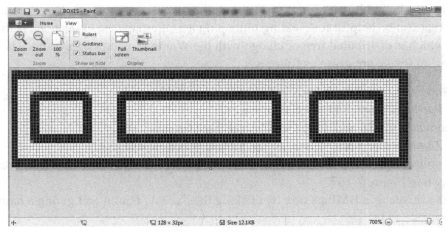

Figure 7.167: The Image Used in This Example

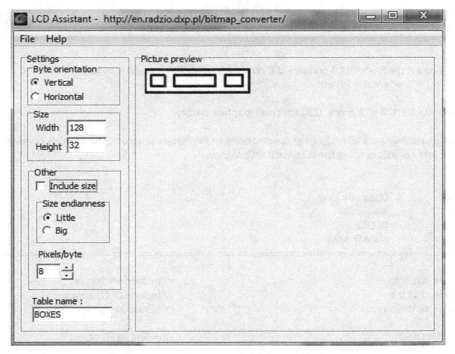

Figure 7.168: Using the LCD Assistant

- Click *File → Save output* and save the file in a folder.
- Start the MPIDE development environment, including the SPI and the OLED libraries. Additionally, include the bitmap array file created above.

The program listing (OLED2) for this project is shown in Figure 7.169. The bitmap image is displayed by calling OLED function **OLED.putBmp** with the size of the image and the bitmap array filename. In this example, the bitmap array file was stored in folder C:\.

Figure 7.170 shows the created image being displayed.

7.32 Project 7.32 – I²C Thermometer With OLED Display

7.32.1 Project Description

In this project, we will be using the PmodTMP2 I²C-based temperature sensor module to measure the ambient temperature and display it on the PmodOLED module in graphical form.

PmodTMP2 is a high-accuracy digital temperature sensor module using the ADT7420 (Analog Devices) temperature sensor chip. The module is based on the I²C bus interface and acts as a slave device. The basic features of the ADT7420 chip are:

- Temperature accuracy of ±0.20°C from −10 to +85°C at 3.0 V
- Temperature accuracy of ±0.25°C from −20 to +105°C from 2.7 to 3.3 V
- Fast temperature conversion (240 ms continuous)

```
*================================================================================
                         CREATING AN IMAGE
                         =================

In this project the PmodOLED graphics LCD module is used. The module is connected to Pmod
connector JPE where the SPI bus is.

PmodOLED is a 128 x 32 pixels, 0.9 inch small graphics display.

The program displays a bitmap image created using the Windows program. The image is converted
Into an array of values using the LCD Assistant program.


Author:        Dogan Ibrahim
Date:          May, 2014
File:          OLED2
Board:         chipKIT MX3
==============================================================================*/
```

```cpp
#include <DSPI.h>                                // include SPI library
#include <OLED.h>                                // include OLED library
#include <c:\boxesout>                           // bitmap array file

OledClass OLED;                                  // define OLED object


void setup()
{
  OLED.begin();                                  // initialize & power ON OLED
}


void loop()
{
  OLED.putBmp(128, 32, BOXES);                   // display image
  OLED.updateDisplay();                          // update display
  delay(5000);                                   // wait 5 seconds
  OLED.clear();                                  // clear display
}
```

Figure 7.169: Program Listing

- Low-power operation (700 μW at 3.3 V operation)
- Overtemperature and undertemperature control pins

The module has 2 × 4 pins for connecting to the Pmod connectors. The module has four selectable addresses.

Figure 7.171 shows a picture of the PmodTMP2 module. The interface connector (J1) pin layout is shown in Table 7.11. The I²C interface uses the two signals SCL and SDA. Module

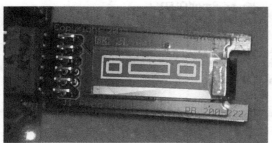

Figure 7.170: The Created Image

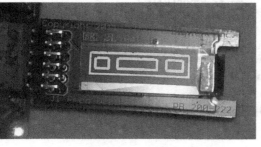

Figure 7.171: PmodTMP2 Module

Table 7.11: PmodTMP2 interface connector.

Connector J1 – I²C Communications		
Pin	**Signal**	**Description**
1, 2	SCL	I²C clock
3, 4	SDA	I²C data
5, 6	GND	Power supply ground
7, 8	VCC	Power supply (3.3 V/5 V)

address is selected using two jumpers (JP1 and JP2) as shown in Table 7.12. By default, the jumpers are not connected and the default module address is 0x4B.

The module provides two open drain outputs for controlling external devices in temperature threshold applications. The module is configured such that when powered up it can be used as a simple temperature sensor without any configuration. At start-up, a 2-byte read without specifying a register will read the value of the temperature from the device, where the first byte will be the most significant byte and the second byte is the least significant byte in the form of 2's complement notation. If the result is shifted right by 3 bits (this is because the 3 least significant bits of the returned data are the event alarm flags for overtemperature and undertemperature conditions) and multiplied by 0.0625 (this is because each LSB of the

Table 7.12: PmodTMP2 module address selection.

JP2	JP1	Address
Open	Open	0×4B (0b1001011)
Open	Shorted	0×4A (0b1001010)
Shorted	Open	0×49 (0b1001001)
Shorted	Shorted	0×48 (0b1001000)

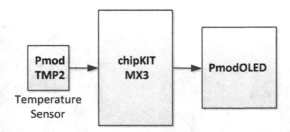

Figure 7.172: Block Diagram of the Project

returned value corresponds to 0.0625°C), then the resulting signed floating point value will be the temperature reading in degrees centigrade.

The block diagram of the project is shown in Figure 7.172.

7.32.2 The I²C Bus

Before going into the details of the project hardware, it is worthwhile to review the basic principles of the I²C bus communications protocol. I²C is a bidirectional two-line communication between a master and one or more slave devices. The two lines are named SDA (serial data) and SCL (serial clock). Both lines must be pulled up to the supply voltage using suitable resistors. Figure 7.173 shows a typical system configuration with one master and three slaves communicating over the I²C bus.

Most high-level language compilers provide libraries for I²C communication. We can also easily develop our own I²C library. Although the available libraries can easily be used, it is worthwhile to look at the basic operating principles of the bus.

The I²C bus must not be busy before data can be sent over the bus. Data is sent serially, and synchronised with the clock. Both SDA and SCL lines are HIGH when the bus is not busy. The START bit is identified by the HIGH-to-LOW transition of the SDA line while the SCL is HIGH. Similarly, a LOW-to-HIGH transition of the SDA line while the SCL is HIGH is identified as the STOP bit. Figure 7.174 shows both the START and STOP bit conditions.

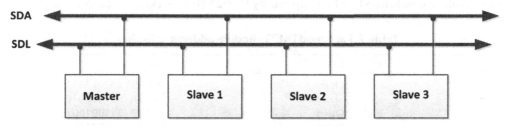

Figure 7.173: I²C System Configuration

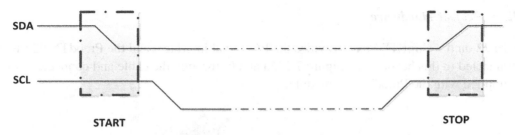

Figure 7.174: START and STOP Bit Conditions

1 bit of data is transferred during each clock pulse. Data on the bus must be stable when SCL is HIGH; otherwise, the data will be interpreted as a control signal. Data can change when the SCL line is LOW. Figure 7.175 shows how bit transfer takes place on the bus.

Each byte of 8 bits on the bus is followed by an acknowledgement cycle. The acknowledgement cycle has the following requirements:

- An addressed slave device must generate an acknowledgement after the reception of each byte from the master.
- A master receiver must acknowledge after the reception of each data byte from the slave (except the last byte).
- The acknowledge signal is identified by a device by lowering the SDA line during the acknowledge clock HIGH pulse.
- A master receiver must signal the end of data to the transmitter by not lowering the SDA line during the acknowledge clock HIGH pulse. In this case, the transmitter leaves the SCL line HIGH so that the master can generate the STOP bit.

The communication over the I²C bus is based on addressing where each device has a unique 8-bit address, usually set up by hardware configuration. Before sending any data, the address of the device that is expected to respond is sent after the START bit.

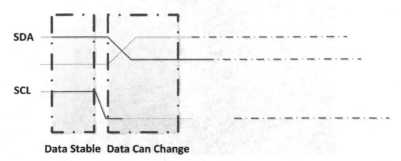

Data Stable Data Can Change

Figure 7.175: Bit Transfer on the Bus

7.32.3 Project Hardware

Header J2 on the chipKIT MX3 provides the I²C signal interface, and the PmodTMP2 must be connected to this header (see Figure 7.176a and b) through the cable and connector assembly supplied with the PmodTMP2 module.

Figure 7.176: (a) The PmodTMP2 module. (b) Connecting the PmodTMP2 Module

Figure 7.177: Project Hardware Setup

Since the PmodTMP2 has open collector outputs, the pull-up resistors on the board must be enabled. Shorting blocks must be installed in jumpers JP1 and JP10 near the Pmod connector JA in order to enable the pull-up resistors. Figure 7.177 shows the hardware setup of the project. The circuit diagram of the project is shown in Figure 7.178.

7.32.4 Project PDL

The chipKIT I²C library called **WIRE** is used to access the I²C signals. This library is normally included in the MPIDE development environment.

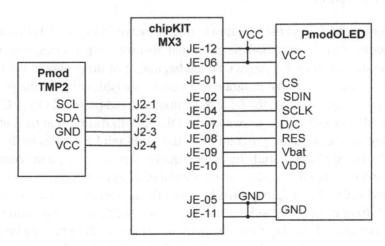

Figure 7.178: Circuit Diagram of the Project

The file **Wire.h** must be included at the beginning of your programs in order to use the I²C interface. The following functions are offered by this library:

Wire.begin(): This function initialises the I²C library and must be called before calling other functions. Optionally the 7-bit slave device address can be specified; otherwise, the interface is a master.

Wire.write(data) or **Wire.write(string)** or **Wire.write(data, length)**: This function sends data over the bus.

Wire.read(): This function reads a byte from the bus.

Wire.beginTransmission(address): This function starts transmission with a slave device whose address is given.

Wire.endTransmission(): This function ends transmission to a slave device that was begun.

Wire.requestFrom(address, quantity): This function is used by the master to request bytes from a slave device. The bytes may then be retrieved with the available() and read() functions. The 7-bit address of the device to request bytes from and the number of bytes requested are arguments of the function.

Wire.available(): This function returns the number of bytes available for retrieval with the read() function.

Wire.onReceive(handler): This function registers a function to be called when a slave device receives a transmission from a master.

Wire.onRequest(handler): This function registers a function to be called when a master requests data from a slave device.

Figure 7.179 shows the PDL of the project.

7.32.5 Project Program

The program is named I²C, and the program listing is shown in Figure 7.180. Although the PmodTMP2 module can measure both negative and positive temperatures, only positive temperatures are displayed in this project. At the beginning of the program, the I²C, SPI, and OLED libraries are included in the program and various variables used in the program are created. Inside the *setup()* routine, the OLED is initialised and powered ON, I²C interface signals are initialised, and *X–Y*-axis is drawn on the OLED with its origin at the bottom left corner of the display. Inside the main program loop, the MSB and LSB bytes of the temperature are read from the PmodTMP2 module by calling the I²C functions **requestFrom** and **receive**. The address of the PmodTMP2 module and the number of bytes requested from the slave device are specified as 0x4B and 2, respectively. Then, the temperature is calculated in degrees centigrade by shifting right 3 bits and multiplying with 0.0625. The temperature is converted into a string and displayed starting from character position (10, 2) of the display. The *Y* pixel value of the temperature is calculated by subtracting the integer value of the temperature from 31.

```
BEGIN
        Include I2C, SPI, OLED libraries
        Start I2C and OLED libraries
        CALLDraw_Axis
        DO FOREVER
                Request 2 bytes from PmodTMP2 module
                Get MSB and LSB bytes
                Calculate the temperature
                Display temperature at (10,2) in numeric format
                Calculate the Y co-ordinate value
                Draw graph of the temperature against time
                IF the end of horizontal axis is reached THEN
                        Clear X value
                        Clear graph
                        CALL Draw_Axis
                ENDIF
                Wait 10 seconds
        ENDDO
END


BEGIN/Draw_Axis
        Clear display
        Draw X-Y axis with the origin at bottom left corner or the display
END/Draw_Axis
```

Figure 7.179: Project PDL

Thus, for example, if the temperature is 10°C, then this will be shown as a pixel at display coordinate $Y = 21$, and so on. Therefore, the display can show temperature values from 0°C (at the bottom of the display, corresponding to actual display coordinate $Y = 31$) to up to 31°C (at the top of the display, corresponding to actual display coordinate $Y = 0$). The time is incremented in steps of 10 s, that is, each horizontal pixel corresponds to 10 s. When the end of the horizontal axis is reached, the X value is cleared, axes are redrawn, and the display continues from the beginning. Notice that during the first iteration, the X value is set to 0 so that display starts from the left-hand side and also the axes are drawn.

A typical output on the display is shown in Figure 7.181.

7.33 Project 7.33 – Using the GPS Module

7.33.1 Project Description

In this project, we will be using the PmodGPS module to get the navigational parameters of our current location on Earth and then to display them on the Serial Monitor.

```
/*=============================================================================*/

                    TEMPERATURE DISPLAY ON OLED
                    ===========================
```

In this project the PmodOLED graphics LCD module is used. The module is connected to Pmod connector JPE where the SPI bus is. In addition, the temperature sensor module PmodTMP2 is connected to connector J2 where the I^{2}C interface signals are. This connection is done using the supplied connectors and through four wires.

PmodOLED is a 128 x 32 pixels, 0.9 inch small graphics display.

The temperature is displayed both in numeric form and also in the form of a graph where the horizontal axis is the time and the vertical axis is the temperature. The display is organized such that each pixel in the vertical direction corresponds to one Degree Centigrade, and in the horizontal direction each pixel corresponds to 10 seconds.

```
Author:        Dogan Ibrahim
Date:          May, 2014
File:          I2C
Board:         chipKIT MX3
=============================================================================*/

#include <Wire.h>                              // include I2C library
#include <DSPI.h>                              // include SPI library
#include <OLED.h>                              // include OLED library

OledClass OLED;                                // define OLED object
unsigned char msb, lsb;
unsignedint temp;
float temperature;
char s[6];
char t, y, x, xold, yold;
charFirst_time = 1;

//
// Clear display and draw an X-Y axis with center at bottom left
//
voidDraw_Axis(void)
{
        OLED.clear();                          // clear display
        OLED.moveTo(0, 31);                    // go to 0,31
        OLED.drawLine(127, 31);                // draw line to 127,31
        OLED.moveTo(0, 31);                    // go to 0,31
        OLED.drawLine(0, 0);                   // draw line to 0,0
        OLED.updateDisplay();                  // update display
}


void setup()
{
        Wire.begin();                          // join I2C bus
        OLED.begin();                          // Power ON OLED
        Draw_Axis();                           // draw the axes
}
```

Figure 7.180: Program Listing *(Continued)*

```
void loop()
{
        Wire.requestFrom(0x4B, 2);              // request 2 bytes from slave device #2
        msb = Wire.receive();                   // receive MSB byte
        lsb = Wire.receive();                   // receive LSB byte

        temp = 256*msb + lsb;                   // store in 16-bits
        temp = temp >> 3;                       // calculate temperature...
        temperature = 0.0625 * temp;            // in Degrees C
        OLED.setCursor(10,2);                   // character cursor at 10,2
        sprintf(s,"%2.2f", temperature);        // convert into string (nn.nn format)
        OLED.putString(s);                      // display temperature on OLED
        t = (int)temperature;                   // get integer part
        y = 31-t;                               // format for the y-axis

        if(First_time == 1)                     // if first time in loop
          {
                x = 0;
                OLED.moveTo(x, y);              // goto 0,y
                First_time = 0;
          }
        else
          {
                OLED.moveTo(xold, yold);        // goto stored positions
                x = x + 1;                      // increment x
                if(x == 127)                    // if end of screen
                  {
                        x = 0;                  // reset x co-ordinate to 0
                        OLED.clear();
                        Draw_Axis();            // re-draw axes
                  }
                else
                        OLED.drawLine(x, y);    // draw line for temperature
          }
        xold = x;                               // save x
        yold = y;                               // save y
        OLED.updateDisplay();                   // update display
        delay(10000);                           // wait 10 seconds
}
```

Figure 7.180: *(cont.)*

The GPS is a satellite-based system that provides location (latitude, longitude, altitude) and time information anywhere on Earth and at any time and in all weather conditions. The system was originally developed for military use but is now freely available to anyone with a GPS receiver.

The GPS system was developed in 1973 by the U.S. Department of Defense, and originally it had 24 satellites orbiting the Earth. The satellites are positioned such that at any time a

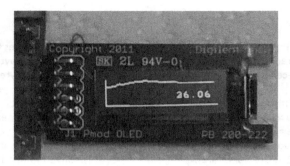

Figure 7.181: Typical Output on the Display

minimum of four satellites are visible from anywhere on Earth. The determination of the latitude and longitude requires information from at least three satellites. The altitude requires a fourth satellite to be available. In practise, up to 10 satellites can be in view in open space.

The GPS system consists of three major segments: space segment, control segment, and user segment. The U.S. Air Force develops, maintains, and operates the space and control segments. Currently the space segment consists of 24–32 satellites in orbit around the Earth. The control segment consists of a master control station and alternate control stations. The user segment consists of the user GPS receivers.

A GPS receiver is a small handheld battery-operated device the size of a mobile phone. It receives signals from the visible satellites and determines the owner's location on Earth with high precision (around 3 m). In general, GPS receivers are composed of an antenna, a processor, and usually a graphical display to show user's navigational parameters. A receiver is often described by its number of channels: this signifies how many satellites it can monitor simultaneously. Currently most low-cost receivers have at least 12 channels.

GPS receivers may additionally include an external input for differential corrections, using the RTCM SC-104 data format. This correction is in the form of an RS-232 serial data at 4800 bits/s. Some receivers have internal correction channels known as Wide Area Augmentation System (WAAS). With this channel, the overall accuracy and integrity of a receiver are enhanced.

Most GPS receivers can send out position data to a PC or any other suitable device using the NMEA 0183 protocol. This protocol basically consists of sentences starting with identifier character "$," followed by a word that identifies the sentence type, and then the navigational data are given, separated with commas. Different GPS receivers send out different types of NMEA 0183 sentences.

In this project, we will be receiving the NMEA 0183 sentences from a PmodGPS module and then display them on the Serial Monitor.

Figure 7.182 shows a picture of the PmodGPS module.

Figure 7.182: PmodGPS Module

The basic features of the PmodGPS module are:

- Six-pin Pmod connector
- Operating voltage 3–3.6 V
- Low power consumption (24–30 mA)
- Built-in antenna (external antenna capability on header J4)
- Horizontal accuracy of 3 m
- Operation with two-wire UART interface
- RTCM SC-104–compatible differential correction capability
- Optional 3 V coin cell battery to reduce the amount of time it takes to acquire the first positional fix

The PmodGPS module operates with serial communication at 9600 Baud (can be increased if required), 8 data bits, no parity bit, and 1 stop bit. The first time the device is powered, it can take up to 2 min to display the positional data. After the first time, acquisition time is reduced to less than 30 s.

Figure 7.183 shows the block diagram of the project.

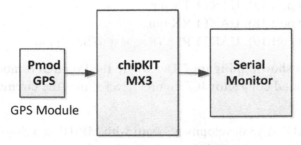

Figure 7.183: Block Diagram of the Project

Table 7.13: PmodGPS connector J1 pin layout.

Pin	Signal	Description
1	3DF	3D-Fix indicator
2	RX	Receive
3	TX	Transmit
4	1PPS	1 pulse per second
5	GND	Power supply ground
6	VCC	Power supply (3.3 V/5 V)

7.33.2 Project Hardware

The PmodGPS is interfaced with the chipKIT MX3 development board through its J1 connector. Table 7.13 shows the J1 pin layout of this connector. The 3DF pin (pin 1) indicates the status of the positional fix, and it stays LOW when a fix is obtained. This pin is toggled every second if a fix cannot be obtained. Also, the green light (LD1) on the board flashes every second if a fix is not obtained. This light turns OFF when a fix is obtained.

Pins 2 and 3 are the serial communication receive and transmit pins, respectively. Pin 4 is the one pulse per second pin (not used in this project). The two-pin connector J2 consists of the Reset and differential correction pins and is not used in this project.

chipKIT MX3 UART interface

The PIC32MX320 microcontroller provides two UART interfaces (UART1 and UART2) with two- or four-wire asynchronous serial interfaces. In this project, the two-wire interface is used with the receive (RX) and transmit (TX) pins. In this project, UART2 interface is used (notice that the Serial Monitor software uses UART1 and using this port for the GPS may interfere with operation of the Serial Monitor).

This interface is accessed from Pmod connector JC (UART1 interface is accessed from JB). The pin configuration of UART2 is:

- JC-01 (logical I/O port 16), UART1 CTS (Clear to Send) pin
- JC-02 (logical I/O port 17), UART1 TX pin
- JC-03 (logical I/O port 18), UART1 RX pin
- JC-04 (logical I/O port 19), UART1 RTS (Request to Send) pin

The hardware setup is shown in Figure 7.184 where the PmodGPS module is connected to the upper row of Pmod connector JC. Figure 7.185 shows the circuit diagram of the project.

When using the chipKIT MX3 development board with MPIDE development environment, the chipKIT serial communications library *SoftwareSerial* can be used.

Figure 7.184: Hardware Setup

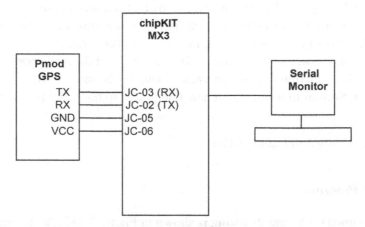

Figure 7.185: Circuit Diagram of the Project

7.33.3 Project PDL

The chipKIT *SoftwareSerial* library provides serial communications capability through any required I/O port of the microcontroller. Because this library is implemented in software, it has the following limitations:

- Only speeds up to 9600 Baud are supported.
- There is no function to check if data is available at the receive buffer.
- The read function is blocking and waits until data becomes available.
- Data received when the receive function is not called is lost, that is, the received data is not stored in a buffer for future use.

```
BEGIN
        Include SoftwareSerial library and create serial object GPS
        Define RX and TX logical I/O pins
        Initialize Serial Monitor to 9600 Baud
        Initialize GPS port to 9600 Baud
        DO FOREVER
                Read data from GPS serial port
                Display read data on Serial Monitor
        ENDDO
END
```

Figure 7.186: Project PDL

The following functions are provided with the *SoftwareSerial* library:

> **SoftwareSerial(RXpin, TXpin)**: This function creates a serial object where the receive (Rxpin) and transmit (Txpin) pins must be specified as arguments to the function.
> **begin(Baud)**: This function sets the communications Baud rate.
> **read()**: This function waits for a character to be available at the serial port and then reads it.
> **print(data)**: This function writes a character (data) to the serial port.
> **println(data)**: Similar to *print()* but a new line is sent to the serial port at the end of the data.

The project PDL is shown in Figure 7.186.

7.33.4 Project Program

The program is named GPS, and its listing is shown in Figure 7.187. At the beginning of the program, **SoftwareSerial** library is included and the logical I/O port numbers of the receiver and transmit pins are defined. The serial object is named **GPS**. Inside the *setup()* routine, the Serial Monitor and the GPS serial port are initialised to operate at 9600 Baud.

The operation of the main program loop is very simple. Here, function **GPS.read** reads a byte from the PmodGPS module. This byte is then sent to the Serial Monitor by calling function **Serial.write**.

Figure 7.188 shows the data received from the PmodGPS module every second. The decoding of some of the sentences is given in Figures 7.189 and 7.190 (further information can be obtained from the PmodGPS product data sheet).

```
/*============================================================================*/

                            USING A GPS
                            ===========

In this project the PmodGPS Global Positioning System (GPS) module is used to get the
navigational parameters of the current location and then to display them on the Serial Monitor.

The PmodGPS module is connected to the upper row of the Pmod connector JPC. The PmodGPS
Module uses the serial communication port pins RX and TX and operates at 9600 ABud.

The program receives the NMEA sentences continuously and displays them on the Serial Monitor.

Author:        Dogan Ibrahim
Date:          May, 2014
File:          GPS
Board:         chipKIT  MX3
============================================================================*/

#include <SoftwareSerial.h>                        // include SoftwareSerial library

#define RXpin 18                                    // RX logical pin number
#define TXpin 17                                    // TX logical pin number

SoftwareSerialGPS(RXpin, TXpin);                    // create serial object GPS

void setup()
{
        Serial.begin(9600);                         // Initialize Serial Monitor
        GPS.begin(9600);                            // Initialize the GPS serial port
}


void loop()
{
        Serial.write(GPS.read());                   // read from the GPS port and display
}                                                   // on the Serial Monitor
```

Figure 7.187: Program Listing

7.34 Project 7.34 – Stepper Motors

7.34.1 Project Description

This project is about using stepper motors in microcontroller-based systems. This is an introductory project where a stepper motor is driven from a microcontroller.

Before going into the details of the project, it is worthwhile to look at the theory and operation of stepper motors briefly.

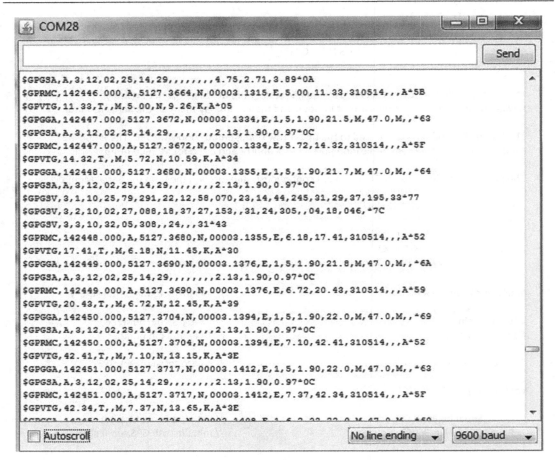

Figure 7.188: Data Received From the PmodGPS Module

Stepper motors are commonly used in printers, disk drives, position control systems, and many more systems where precise position control is required. Stepper motors come in a variety of sizes, shapes, strengths, and precision. There are two basic types of stepper motors: unipolar and bipolar.

Unipolar stepper motors

Unipolar stepper motors have two identical and independent coils with centre taps, and have five, six, or eight wires (see Figure 7.191).

Unipolar stepper motors can be driven in three modes: one-phase full-step sequencing, two-phase full-step sequencing, and two-phase half-step sequencing.

$GPGGA,142448.000,5127.3680,N,00003.1355,E,1,5,1.90,21.7,M,47.0,M,,*64

142448.000	UTC Time (hhmmss.sss)
5127.3680	Latitude (ddmm.mmmm)
N	North/South Indicator
00003.1355	Longitude (dddmm.mmmm)
E	East/West Indicator
1	Position Fix Indicator
5	Satellites Used
1.90	HDOP
21.7	MSL Altitude
M	Units
47.0	Geoidal Separation
M	Units
*64	Checksum

Figure 7.189: Decoding the $GPGGA Sentence

$GPRMC,142451.000,A,5127.3717,N,00003.1412,E,7.37,42.34,310514,,,A*5F

142451.000	UTC time (hhmmss.sss)
A	Status (A = Data Valid)
5127.3717	Latitude (ddmm.mmmm)
N	North/South Indicator
00003.1412	Longitude (dddmm.mmmm)
E	East/West Indicator
7.37	Speed Over Ground (Knots)
42.34	Course Over Ground (Degrees)
310514	Date (ddmmyy)
A	Mode
5F	Checksum

Figure 7.190: Decoding the $GPRMC Sentence

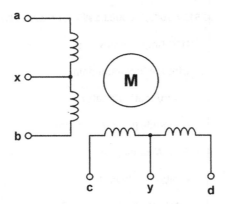

Figure 7.191: Unipolar Stepper Motor Windings

One-phase full-step sequencing

Table 7.14 shows the sequence of sending pulses to the motor. Each cycle consists of four pulses.

Two-phase full-step sequencing

Table 7.15 shows the sequence of sending pulses to the motor. The torque produced is higher in this mode of operation.

Two-phase half-step sequencing

Table 7.16 shows the sequence of sending pulses to the motor. This mode of operation gives more accurate control of the motor rotation, but requires twice as many pulses for each cycle.

Table 7.14: One-phase full-step sequencing.

Step	a	c	b	d
1	1	0	0	0
2	0	1	0	0
3	0	0	1	0
4	0	0	0	1

Table 7.15: Two-phase full-step sequencing.

Step	a	c	b	d
1	1	0	0	1
2	1	1	0	0
3	0	1	1	0
4	0	0	1	1

Table 7.16: Two-phase half-step sequencing.

Step	a	c	b	d
1	1	0	0	0
2	1	1	0	0
3	0	1	0	0
4	0	1	1	0
5	0	0	1	0
6	0	0	1	1
7	0	0	0	1
8	1	0	0	1

The motor can be connected to a microcontroller using power transistors or power MOSFET transistors.

Bipolar stepper motors

Bipolar stepper motors have two identical and independent coils and four wires, as shown in Figure 7.192.

The control of bipolar stepper motors is slightly more complex. Table 7.17 shows the typical driving sequence. The "1" and "0" denote the logic levels applied to the motor legs. Bipolar stepper motors are usually driven using H-bridge circuits.

The bipolar consists of two coils, but there is no centre tap. As a result of this, the bipolar motor requires a controller where the current flow through the coils can be reversed. A bipolar motor is capable of higher torque since entire coils may be energised, not just half of the coils. Bipolar stepper motors are usually controlled using H-bridge circuits where the current flow through the coils can easily be reversed.

In this project, a bipolar stepper motor is used. The motor is rotated 10 revolutions in one direction, then stopped for 5 s, and then rotated 10 revolutions in the other direc-

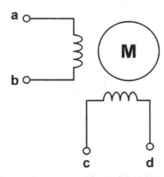

Figure 7.192: Bipolar Stepper Motor Windings

Table 7.17: Bipolar stepper motor driving sequence.

Step	a	c	b	d
1	1	1	0	0
2	0	1	1	0
3	0	0	1	1
4	1	0	0	1

tion, and is then stopped. The PmodSTEP stepper motor controller module is used in this project.

Figure 7.193 shows the block diagram of the project.

7.34.2 Project Hardware

Figure 7.194 shows a picture of the PmodSTEP stepper motor controller module. This module is based on the L293D, which is a quadruple half-H motor controller chip. The Pmod-STEP has the following features:

- External power for larger motors
- Signal status and power LEDs
- Compatibility with Pmod connectors

The L293D motor controller chip has the following features:

- 4.5–36 V supply voltage
- Up to 600 mA current per channel (peak 1.2 A)
- Output clamp diodes for transient suppression
- TTL-compatible logic inputs
- Easy control via *Enable* inputs

The L293D chip contains four high-current driver inputs named 1A–4A, with corresponding outputs named as 1Y–4Y. As shown in the logic diagram in Figure 7.195, 1A and 1B drivers are enabled using the 1,2EN pins. Similarly, 3A and 4A drivers are enabled using the 3,4EN logic pins.

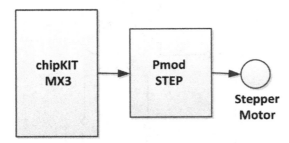

Figure 7.193: Block Diagram of the Project

Figure 7.194: The PmodSTEP Module

PmodSTEP module is connected to the development board through Pmod connector J1. This connector has connections to the L293D chip with the following pin configuration:

Pin Number	Function
1	SIG1
2	SIG2
3	SIG3
4	SIG4
5	GND
6	VCC
7	SIG5 (input 1A)
8	SIG6 (input 2A)
9	SIG7 (input 3A)
10	SIG8 (input 4A)
11	GND
12	VCC

All the J1 signals (SIG1–SIG8) drive LEDs through MOSFET switching transistors so that the state of each pin can be observed. The outputs of the L293D chip are brought to a four-pin header called J2 with the following pin configuration:

Pin Number	Function
1	SM1 (output 1Y)
2	SM2 (output 2Y)
3	SM3 (output 3Y)
4	SM4 (output 4Y)

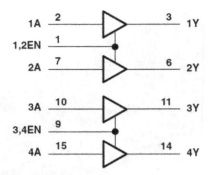

Figure 7.195: Logic Diagram of the L293D Chip

External power to the PmodSTEP module can be selected via the JP1 jumper. In this project, the internal +5 V is used to provide power to the motor and also to the controller module.

The PmodSTEP module is connected to Pmod connector JA, and the JA power jumper is set to +5 V so that +5 V power is applied to the module. PmodSTEP jumper JP1 is set to VCC (left position) so that power is taken from the VCC and not from an external source.

The interface between the PmodSTEP module and the development board is thus as follows:

JA Pin Number	Logical I/O Port Number	Port	PmodSTEP Pin
JA-01	0	RE0	SIG1 (not used)
JA-02	1	RE1	SIG2 (not used)
JA-03	2	RE2	SIG3 (not used)
JA-04	3	RE3	SIG4 (not used)
JA-07	4	RE4	SIG5 (input 1A)
JA-08	5	RE5	SIG6 (input 2A)
JA-09	6	RE6	SIG7 (input 3A)
JA-10	7	RE7	SIG8 (input 4A)

The type of motor used in this project is the 39HS02 (see Figure 7.196). This motor has the following features:

- Bipolar stepper motor
- 1.8° step angle (200 steps for a complete revolution)
- ±5% step angle accuracy
- 0.6 A phase current
- Four leads (coil 1: brown + grey, coil 2: orange + green)

Figure 7.197 shows the hardware setup of the project. The circuit diagram of the project is shown in Figure 7.198a. Figure 7.198b shows the Pmod connection to the development board and the motor. Notice that the motor coils are connected to microcontroller port pins RE4–RE7. Letters **a–d** refer to motor coil connections (see Figure 7.192).

Figure 7.196: 39HS02 Bipolar Stepper Motor

7.34.3 Project PDL

The project PDL is shown in Figure 7.199. At the beginning, the I/O pins connected to the motor are configured as outputs. Inside the main program loop, the motor is rotated three revolutions in one direction, then stopped for 5 s, and then rotated three revolutions in the other direction, and is then stopped.

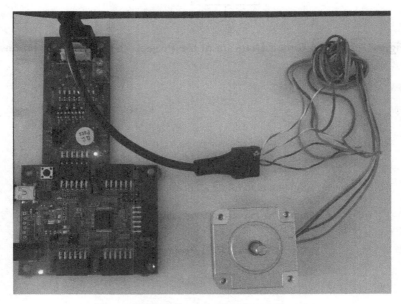

Figure 7.197: Project Hardware Setup

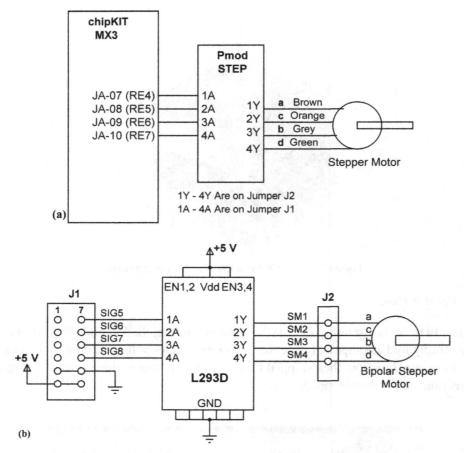

Figure 7.198: (a) Circuit Diagram of the Project. (b) Pmod Connections

```
BEGIN
        Define motor connections a,b,c,d
        Define required number of pulses for 10 revolutions
        Configure motor connections as outputs
        Send pulses to rotate 10 revolutions in one direction
        Wait 5 seconds
        Send pulses to rotate 10 revolutions in other direction
END
```

Figure 7.199: Project PDL

```
/*===========================================================================*/

                        BIPOLAR STEPPER MOTOR CONTROL
                        ==============================
```

In this project the PmodSTEP motor control module is used to control a bipolar stepper
motor. The module is connected to Pmod connector JA of the chipKIT MX3 development board.

The 39HS02 bipolar stepper motor is used in this project. This motor has step angle of 1.8
degrees. Thus, if REV is the required number of revolutions then 200*REV/4 pulses must be sent
to the motor. For 3 revolutions, we have to send 200*3/4 = 150 pulses.

The stepper motor is controlled as follows:

The motor is rotated 3 revolutions in one direction, then stopped for 5 seconds, and then rotated
3 revolutions in the other direction, and is then stopped.

The stepper motor connections are:
Coil 1: a, b
Coil 2: c, d

```
Author:       Dogan Ibrahim
Date:         June, 2014
File:         MOTOR
Board:        chipKIT MX3
===========================================================================*/
```

```c
unsigned char REVS;
unsigned inti ,pulses;

//
// This function rotates the motor in forward direction
//
void Forward()
{
        PORTE = 0x30;
        delay(5);
        PORTE = 0x60;
        delay(5);
        PORTE = 0xC0;
        delay(5);
        PORTE = 0x90;
        delay(5);
}


//
// This function rotates the motor in reverse direction
//
void Reverse()
{
        PORTE = 0x90;
        delay(5);
        PORTE = 0xC0;
        delay(5);
        PORTE = 0x60;
        delay(5);
```

Figure 7.200: Program Listing *(Continued)*

```
            PORTE = 0x30;
            delay(5);
    }


    void setup()
    {
            REVS = 3;                           // required number of revolutions
            pulses = 200*REVS/4;                // pulses to be sent
            TRISE=0;                            // configure PORTE as output
    }



    void loop()
    {
    //
    // First rotate clockwise 3 revolutions
    //
            for(i = 0; i< pulses; i++)
                    Forward();
            delay(5000);                        // wait 5 seconds
    //
    // Now rotate anticlockwise 10 revolutions
    //
            for(i = 0; i< pulses; i++)
                    Reverse();
            while(1);                           // end. Wait here forever
    }
```

Figure 7.200: *(cont.)*

7.34.4 Project Program

The program is called MOTOR, and its listing is shown in Figure 7.200. Inside the *setup()* routine, the required number of revolutions and number of pulses required to obtain these revolutions are specified. Since the motor stepping angle is 1.8°, the number of pulses to be sent to the motor to make the motor turn REV revolutions is calculated as $(200 \times REV)/4$. Two functions are used to rotate the motor. Function **Forward** sends pulses (see Table 7.17) to PORTE in the following order to rotate the motor forwards. A small delay (5 ms) is used between each pulse. This delay determines the speed of rotation:

7	6	5	4	3	2	1	0	
RE7	RE6	RE5	RE4	RE3	RE2	RE1	RE0	
d	b	c	a					
0	0	1	1	0	0	0	0	0x30
0	1	1	0	0	0	0	0	0x60
1	1	0	0	0	0	0	0	0xC0
1	0	0	1	0	0	0	0	0x90

In order to rotate the motor in the opposite direction, the pulse order must be reversed. Function **Reverse** sends pulses (see Table 7.17) to PORTE in the following order to rotate in the reverse direction. A small delay (5 ms) is used between each pulse. This delay determines the speed of rotation:

7	6	5	4	3	2	1	0	
RE7	RE6	RE5	RE4	RE3	RE2	RE1	RE0	
d	b	c	a					
1	0	0	1	0	0	0	0	0x90
1	1	0	0	0	0	0	0	0xC0
0	1	1	0	0	0	0	0	0x60
0	0	1	1	0	0	0	0	0x30

You should see the four LEDs E–H flashing as the motor is rotating.

Using the MPLAB IDE with the chipKIT Pro MX7 Development Board

MPLAB IDE® is Microchip's popular integrated development environment. In this appendix, the steps for using the chipKIT Pro MX7 board with MPLAB IDE and Microchip's C32 compiler are shown. The chipKIT MX3 development board cannot be used directly with the MPLAB IDE, and additional programming hardware (e.g., Microchip PICkit 3) is required. chipKIT Pro MX7 is however, a more advanced development board and can be used directly with the MPLAB IDE.

In this appendix, a simple program example is given that flashes LED 1 on the chipKIT Pro MX7 development board using the MPLAB IDE and the C32 compiler. Readers who wish to carry out their developments using the MPLAB IDE environment should find this appendix useful.

Figure A.1 shows a picture of the chipKIT Pro MX7 development board. The basic features of this board are:

- PIC32MX795F512L microcontroller
- 10/100 Ethernet
- 2 × CAN network interfaces
- 4 × LEDs
- 3 × push-button switches
- 2 × USB connectors
- 2 × I²C bus interfaces
- MPLAB IDE and MPIDE support
- Program/debug circuit
- 6 × Pmod connectors

In this project, a program is developed using the MPLAB C32 compiler to flash LED 1 on the development board. This LED is connected to bit 12 of PORTG (RG12) and has the logical I/O port number 51. The program listing is shown in Figure A.2.

The steps to compile and load the program to the program memory of the chipKIT Pro MX7 development board are given as follows:

Step 1: Start the MPLAB IDE (see Figure A.3).
Step 2: Click *Project → Project Wizard* and click *Next*.

Figure A.1: chipKIT Pro MX7 Development Board

Select the device type as PIC32MX795F512L, (see Figure A.4), and click *Next*.

Step 3: Select *Microchip PIC32 C-compiler Toolsuite* and click *Next* (see Figure A.5).

Step 4: Select the source file : LEDF.C (see Figure A.6) and click *Next*.

Step 5: Add source file to the project by selecting it in the left pane and then clicking *Add*, and then click *Next* (see Figure A.7).

Step 6: Click *Finish* to exit the project wizard (see Figure A.8).

Step 7: You should see a list of project folders and the source files you have added to the project. Double click source file LEDF.C to display its contents (see Figure A.9).

Step 8: Click *Debugger → Select Tool → Licensed Debugger*.

Connect the chipKIT MX7 development board to the PC using a USB-A to micro-B cable, connected to micro USB connector called DEBUG (connector J15), positioned on the left hand side of the board just below the Ethernet connector. Make sure that the jumper J3 is connected to position DBG.

When the licensed debugger is selected as the programming or debugging device, the MPLAB IDE will check the version number of the firmware running on the debugger and offer to update if it is out of date with the version of MPLAB being used.

Step 9: Check the Configuration Bits. Click *Configure → Configuration Bits* from the top drop down menu of the IDE. Unclick the box "Configuration Bits set in code" so that the values given here take effect. Change the following Configuration Bits by clicking on the Bit to be changed (see Figure A.10):

PLL Input Divider:	12
PLL Multiplier:	24
System PLL Output Clock Divider:	256
Oscillator Selection Bits:	Primary osc w/PLL (XT+,HS+,EC+PLL)
Primary Oscillator Configuration:	XT osc mode
Peripheral Clock Divisor:	Pb_Clk is Sys_Clk/1
Watchdog Timer Enable:	WDT Disabled

```
/*******************************************************************************
                    chipKIT Pro MX7 FLASHING LED
                    ============================

This is a very simple MPLAB PIC32 program which shows how to use the MPLAB IDE with
The chipKIT Pro MX7 development board.

In this program LED1 on the board is flashed. The logical I/O port number of this LED is 51.
This is pin 12 of PORTG (RG12).

Author:        Dogan Ibrahim
Date:          June, 2014
File:          chipKITProMX7
*******************************************************************************/
//
// Adds support for PIC32 Peripheral library functions and macros
//
#include <plib.h>
#pragma config ICESEL = ICS_PGx1

//
// Microcontroller clock
//
#define SYS_FREQ  (80000000)


voidSmall_delay()
{
        int j, k;
        for(k = 0; k < 10000; k++)j++;
}


//
// Start of main program
//
int main(void)
{

//
// Configure RD0 as output
//
        PORTSetPinsDigitalOut(IOPORT_G, BIT_12);

//
// Turn OFF the LED to start with
//
        PORTClearBits(IOPORT_G, BIT_12);


        for(;;)
        {
                PORTSetBits(IOPORT_G, BIT_12);           // turn ON LED1
                Small_delay();                           // small delay
                PORTClearBits(IOPORT_G, BIT_12);         // turn OFF LED  1
                Small_delay();                           // small delay
        }
}
```

Figure A.2: Program Listing

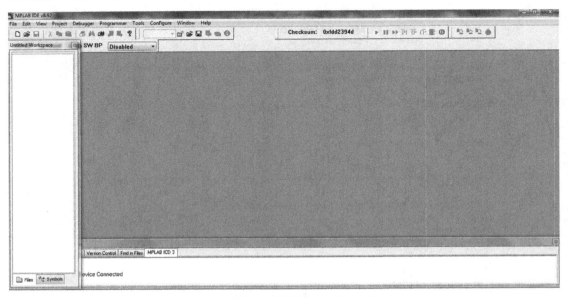

Figure A.3: MPLAB IDE Start-Up Window

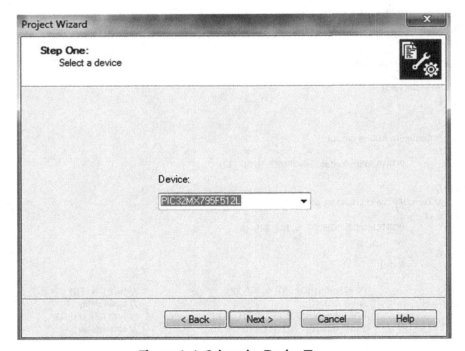

Figure A.4: Select the Device Type

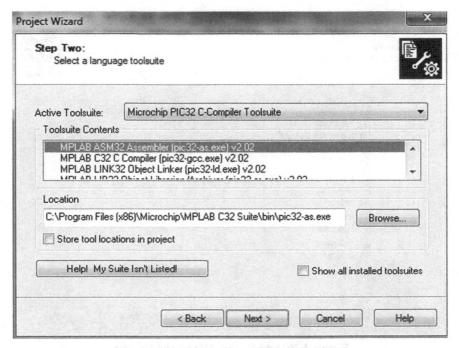

Figure A.5: Select the Compiler

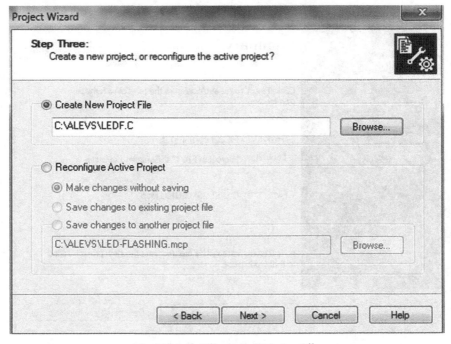

Figure A.6: Select the Source File

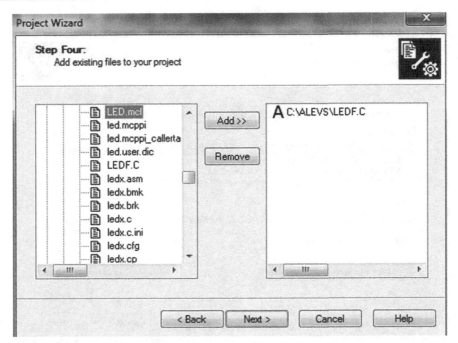

Figure A.7: Add Source File to the Project

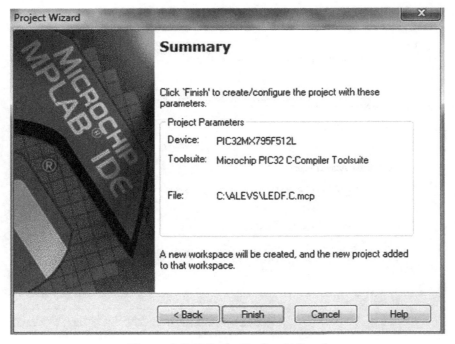

Figure A.8: Exit the Project Wizard

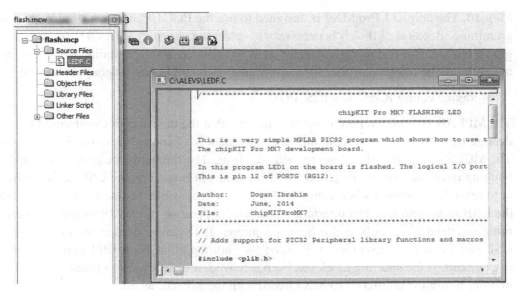

Figure A.9: File LEDF.C

Address	Value	Field	Category	Setting
		☐ Configuration Bits set in code.		
1FC0_2FF0	FFFFFFFF	USERID		
		FSRSSEL	SRS Select	SRS Priority 7
		FMIIEN	Ethernet RMII/MII Enable	MII Enabled
		FETHIO	Ethernet I/O Pin Select	Default Ethernet I/O
		FCANIO	CAN I/O Pin Select	Default CAN I/O
		FUSBIDIO	USB USID Selection	Controlled by the USB Module
		FVBUSONIO	USB VBUS ON Selection	Controlled by USB Module
1FC0_2FF4	FFFFFFFF	FPLLIDIV	PLL Input Divider	12x Divider
		FPLLMUL	PLL Multiplier	24x Multiplier
		UPLLIDIV	USB PLL Input Divider	12x Divider
		UPLLEN	USB PLL Enable	Disabled and Bypassed
		FPLLODIV	System PLL Output Clock Divider	PLL Divide by 256
1FC0_2FF8	FF7FCFFB	FNOSC	Oscillator Selection Bits	Primary Osc w/PLL (XT+,HS+,EC+PLL)
		FSOSCEN	Secondary Oscillator Enable	Enabled
		IESO	Internal/External Switch Over	Enabled
		POSCMOD	Primary Oscillator Configuration	Primary osc disabled
		OSCIOFNC	CLKO Output Signal Active on the OSCO Pin	Disabled
		FPBDIV	Peripheral Clock Divisor	Pb Clk is Sys Clk/1
		FCKSM	Clock Switching and Monitor Selection	Clock Switch Disable, FSCM Disabled
		WDTPS	Watchdog Timer Postscaler	1:1048576
		FWDTEN	Watchdog Timer Enable	WDT Disabled (SWDTEN Bit Controls)
1FC0_2FFC	7FFFFFFF	DEBUG	Background Debugger Enable	Debugger is disabled
		ICESEL	ICE/ICD Comm Channel Select	ICE EMUC2/EMUD2 pins shared with PGC2/PGD2

Figure A.10: Configuration Bits

Step 10: The chipKIT Pro MX7 is designed to use the PGC1/PGD1 pair of pins for programming. Because of this, it is necessary to select the use of PGC1/PGD1 for the debugging interface. The following statement must be included at the beginning of the program to configure the microcontroller for use with the on-board licensed debugger circuit:

#pragma config ICESEL = ICS_PGx1

The MPLAB IDE may report an error indicating that the device is not configured for debugging until a program containing this statement has been programmed into the board. The MCLR pin on the PIC32 microcontroller is used by the hardware programming/debugging interface to reset the processor. This same pin is used by the USB serial converter to reset the processor when using the MPIDE. It is possible that the reset function from the USB serial interface can interfere with correct operation of the Microchip programming and debugging tools. If this happens, jumper JP11 can be used to disconnect the USB serial converter reset circuit. Remove the shorting block from JP11 to disable the reset circuit. If the shorting block has been removed, it is necessary to reinstall it on JP11 in order to use the chipKIT Pro MX7 board with the MPIDE again.

Step 11: Click *Build All* from the top drop down menu to compile the program; you should have no errors (See Figure A.2).

Step 12: Use the *Debugger → Run* command to program all memories on the device and start to run the program on the target microcontroller in debug mode.

The LED should now start flashing.

Reloading the BootLoader

Using the Microchip development tools to program the chipKIT Pro MX7 will erase the BootLoader loaded at the factory. To use the board with the chipKIT MPIDE again, it is necessary to program the BootLoader back onto the board. The programming file for the BootLoader programmed into the board by Digilent at the factory is available for download from the product page for the chipKIT Pro MX7 on the Digilent Inc. web site (in addition, the BootLoader source code is available in the chipKIT project repository at web site: www.github.com/chipKIT32/pic32-Arduino-BootLoader).

To reprogram the BootLoader **using MPLAB**, perform the following steps:

- Use the *Configure → Select Device* menu to select the PIC32MX795F512L.
- Use the *Debugger → Select Tool → Licensed Debugger*.
- Use the *File Import* dialog box to navigate to and select the BootLoader programming downloaded from the Digilent web site. The file name will be something like: chipKIT_ BootLoader_MX7cK.hex.
- Use the *Programmer → Program* command to program all memories on the device.

Index

Printed in the United States
By Bookmasters